Living with Darwin

PHILOSOPHY IN ACTION

Small Books about Big Ideas

WALTER SINNOTT-ARMSTRONG, SERIES EDITOR

This new series publishes short, accessible, lively, and original books by prominent contemporary philosophers. Using the powerful tools of philosophical reasoning, the authors take on our most pressing and difficult questions—from the complex personal choices faced by ordinary individuals in their everyday lives to the major social controversies that define our time. They ultimately show the essential role that philosophy can play in making us think, and think again, about our most fundamental assumptions.

LIVING WITH DARWIN

Evolution, Design, and the Future of Faith

PHILIP KITCHER

OXFORD
UNIVERSITY PRESS
2007

OXFORD
UNIVERSITY PRESS

Oxford University Press, Inc., publishes works that further
Oxford University's objective of excellence
in research, scholarship, and education.

Oxford New York
Auckland Cape Town Dar es Salaam Hong Kong Karachi
Kuala Lumpur Madrid Melbourne Mexico City Nairobi
New Delhi Shanghai Taipei Toronto

With offices in
Argentina Austria Brazil Chile Czech Republic France Greece
Guatemala Hungary Italy Japan Poland Portugal Singapore
South Korea Switzerland Thailand Turkey Ukraine Vietnam

Published by Oxford University Press, Inc.
198 Madison Avenue, New York, New York 10016
www.oup.com

Library of Congress Cataloging-in-Publication Data

Kitcher, Philip, 1947–
Living with Darwin : evolution, design, and the future of faith / Philip Kitcher.
p. cm. — (Philosophy in action)
Includes bibliographical references and index.
ISBN-13: 978-0-19-531444-1
1. Evolution (Biology) 2. Intelligent design (Teleology) 3. Religion
and science. I. Title.
QH366.2.K58 2007
576.8—dc22
2006049819

5 7 9 8 6 4

Printed in the United States of America
on acid-free paper

For Steffi, and in
memory of David

CONTENTS

PREFACE

It began with an issue of *TV Guide*. In 1979, shortly after the birth of our elder son, Andrew, I was searching for some entertainment sufficiently undemanding to soothe two new parents, when my eye fell on a glossy inset, advertising a free book that would set its readers straight on the crucial question of "origins." Intrigued, I filled out the attached form, and, in due course, received my own copy of a creationist classic, *The Remarkable Birth of Planet Earth*, whose author, Henry Morris, was then the head of the Institute for Creation Research.

During the 1970s, I had become increasingly interested in philosophical aspects of the life sciences, and, by the time I read Morris' slim volume, I had already studied a significant amount of evolutionary biology. To my chagrin, however, I discovered that there were objections to Darwinism to which I couldn't immediately formulate a convincing answer. Worse still, I knew that many of my friends, with dimmer memories of high school or college science, would be even less well equipped to respond to creationist rhetoric. I imagined them, frustrated and tongue-tied, as they tried to convince their local school boards that evolution belonged in the curriculum, and belonged without any balance from "creation science." I resolved to write something that would help.

About eighteen months later, at a lunch with two friends, Harry and Betty Stanton, who had begun a new publishing venture, Bradford Books, I found myself buoyed by their enthusiasm—and promised to write a book defending Darwin against the creationist challenge. With enormous help from Patricia Kitcher, coauthor of the book's final chapter and virtually a coauthor of the whole, *Abusing Science* was quickly written and quickly published. It joined other defenses by Niles Eldredge, Douglas Futuyma, and Michael Ruse, and I like to think our combined efforts did something to provide the American public with a more adequate account of the credentials of evolutionary theory and the shortcomings of "creation science."

During the early 1980s, as there were periodic agitations by zealous creationists, I returned to the fray, engaging in public debates, either one-on-one or with allies (including, on a memorable occasion, the redoubtable Kenneth Miller). Gradually, however, the books, the articles, the debates, and, perhaps most importantly, a coordinated system of support for science teachers had an impact on evangelical creationism. By the late 1980s, it appeared that the movement had run out of steam.

We had scotch'd the snake, not killed it. A decade later, there were rumblings of a new form of opposition to Darwin, "intelligent design theory." At first, they seemed minor variations on old themes, but, by the early years of the twenty-first century, it was plain that the new packaging had convinced a very large number of American citizens—including their president. Friends began to suggest, gently, that I might take up the cudgels again.

As I pondered their urgings, I began to realize that I wanted to write a rather different book from the one I had published in 1982. Not only did I want to respond to the fallacious arguments of the spokesmen for intelligent design, but I also hoped to make clear the structure of the powerful case for Darwinian evolutionary theory, to formulate it in a way that people with no great training in science, history, or philosophy could appreciate. Beyond that, I also aimed to respond to the concerns of the thoughtful people who are beguiled by the advertisements for intelligent design, to expose just what it is that is threatening about Darwinism, and to point to the deeper issues that underlie this recurrent conflict. So I have written in the conviction that we shall not escape the cycle of controversy until it is completely clear what lies at the bottom of it all.

Over a quarter of a century, I have acquired many debts. To my great regret, I can no longer thank my friend and mentor, Stephen Jay Gould, tireless defender of Darwin, who extended his hospitality when I wrote *Abusing Science*, and who supported, criticized, and encouraged many of my subsequent writings. From Steve, from Dick Lewontin, and from the late Ernst Mayr, I have learned much about evolution and about Darwinian ideas. While I was preparing to debate creationists in the 1980s, I received staunch support and instruction from Philip Riegle, Bob Schadewald, and especially Malcolm Kottler. For the work of Stanley Weinberg, founder of the Committees of Correspondence, and that of Eugenie Scott, director of the

National Center for Science Education, I, like so many others who have worked on behalf of the public understanding of evolutionary theory, can only express profound admiration and gratitude.

After I wrote a draft of the present book, I circulated it among a significant number of friends, who, with amazing speed, provided me with detailed and informative comments that have greatly improved it. On evolutionary issues, I have been enlightened enormously by the insights and questions of Jerry Coyne, Dick Lewontin, and Bill Loomis. Marty Chalfie, Stuart Firestein, Darcy Kelley, and Bob Pollack have helped me with a wide range of biological questions. Dave Walker has aided me on matters of geology, while David Helfand and Erick Weinberg have given me the perspectives from astrophysics and physics. My fellow philosophers have been most generous with advice and suggestions. I am grateful to David Albert, John Dupré, Jim Lennox, Isaac Levi, Chris Peacocke, Joseph Raz, Alex Rosenberg, Elliott Sober, and Michael Weisberg.

With respect to the religious issues, I have gained much from the comments of Stephen Grover, Martha Nussbaum, Allan Silver, and, especially, Wayne Proudfoot. Kent Greenawalt has shared with me his enormous expertise about constitutional issues and the relations between church and state. David Cohen, Stephanie Lewis, Clyde and Cynthia Rettig, Michael Rothschild, Jim Shapiro, and Jonathan Weiner have all helped me clarify my ideas and make them more accessible.

That all these immensely talented and very busy people provided me with such a wealth of good advice says much about their generosity—and perhaps something about the seriousness with which many educated people take the task of informing a broader public.

After the manuscript was accepted, Peter Ohlin, the philosophy editor for Oxford University Press, suggested that it might appear in a new series edited by Walter Sinnott-Armstrong. In consequence, the final version has benefited from Walter's extensive and constructive comments. I am grateful to him, and to Peter for his judicious advice and strong support. I would also like to thank Robert Miller and Latha Menon of Oxford University Press for some helpful suggestions. I am also grateful to Paula Cooper for her thorough and sensitive editing of my prose.

Three other readers who advised me deserve special thanks. As in 1981, Patricia Kitcher offered a fund of wise counsel and good suggestions. In *Abusing Science*, our sons, Andrew and Charles, figured only as illustrations of some basic genetics—although Andrew, then two, made an unforgettable point about academic overinvolvement one evening, when, a pile of scribbled pages before him, he announced that he would come to dinner once he had "finished his chapter." Now a medical student, Andrew has given me the benefit of a detailed biological knowledge far greater than my own. Charles, who, as he leaves law school, has published his own ideas about how courts should respond to curricular challenges, has given me equally valuable advice from a legal and social perspective.[1] It has really been a family effort.

A final memory. In the spring of 2001, I gave a lecture at Columbia on relations between science and religion. One of the most distinguished American philosophers of recent years, David Lewis, attended, and he and Steffi Lewis stayed for the dinner that followed. David and I talked that evening about a number of issues I raised in my lecture, topics to which we returned in several phone conversations in the succeeding months. But I never saw David again. His death, in September 2001, deprived the philosophical world of a towering intellect. I lost a good friend, half of a wonderful couple, who have cheered, supported, and enlightened Pat and me for over 30 years. This book is dedicated to David and Steffi in great admiration and appreciation.

Living with Darwin

The Sea of Faith
Was once, too, at the full, and round earth's shore
Lay like the folds of a bright girdle furl'd.
But now I only hear
Its melancholy, long, withdrawing roar,
Retreating, to the breath
Of the night-wind, down the vast edges drear
And naked shingles of the world.

Ah, love, let us be true
To one another! for the world, which seems
To lie before us like a land of dreams,
So various, so beautiful, so new,
Hath really neither joy, nor love, nor light,
Nor certitude, nor peace, nor help for pain;
And we are here as on a darkling plain
Swept with confused alarms of struggle and flight,
Where ignorant armies clash by night.

Matthew Arnold, *Dover Beach*

Philosophy alone can boast (and perhaps it is no more than the boast of philosophy) that her gentle hand is able to eradicate from the human mind the latent and deadly principle of fanaticism.

Edward Gibbon, *Decline and Fall of the Roman Empire*

Chapter One

DISINTERRING DARWIN

In 1882, when Charles Darwin died, his family planned to bury him in the churchyard of the village of Down in Kent, where, in his retreat from the bustle of scientific debate, he had spent the last 40 years of his life. Their private plans were overridden by a public campaign, orchestrated by those who had championed Darwin's ideas, and it was decided quickly that he should be interred in Westminster Abbey, among the other luminaries of British science. Whatever doubts they may have harbored, leading figures of the church declared their satisfaction that "properly understood," the evolutionary ideas that had seemed so inflammatory in the 1860s, were perfectly compatible with Christian doctrine. Darwin's own agnosticism, well concealed by his cautious attempts to avoid alienating potential readers—as well as to ensure that the religious sensibilities of his wife Emma (née Wedgwood) were not offended by his expression of his ideas—went unmentioned. Instead, the many tributes from the pulpit heralded him as an old-fashioned Deist, perhaps even as an interpreter of God's

Book of Nature. As one of the eulogies put it, "This man, on whom years of bigotry and ignorance poured out their scorn, has been called a materialist. I do not see in all his writings one trace of materialism. I read in every line the healthy, noble, well-balanced wonder of a spirit profoundly reverent, kindled into deepest admiration for the works of God." With words like these, Darwin was laid in his place of honor beneath the monument to Newton. It seemed as though the church had made its peace with him.

Peace would not last, of course. Almost exactly a century after Darwin was acclaimed as a "spirit profoundly reverent," his detractors petitioned in American courts to protect innocent schoolchildren from the corrosive influence of his theory. In many parts of the world, Scandinavia, Australia, Korea, the Netherlands, the years since 1882 have been punctuated by periodic attempts to disinter Darwin, to repudiate the soothing rhetoric that accompanied his burial, and to expose him as a ruthless enemy of right religion. Nowhere have these efforts been more strenuous than in the United States, where defenders of evangelical Christianity have campaigned in the 1920s, in the 1970s and 1980s, and again today, to remove his ideas from science classrooms or to "balance" them with theories more friendly to faith. Current discussions renew many of the arguments that were traded in the Tennessee courtroom where, in the summer of 1925, John Scopes was arraigned for teaching the "monkey theory." These arguments were joined with less flair in 1982, when Darwinism was again called to the bar in neighboring Arkansas. Through all these episodes, leaders of the anti-evolutionary movement have been consistently clear that Darwin has had a dreadful influence on subsequent culture.

"Evolution is the root of atheism, of communism, Nazism, behaviorism, racism, economic imperialism, militarism, libertinism, anarchism, and all manner of anti-Christian systems of belief and practice."[2] Initially, when the question "which nineteenth-century thinker has had the most damaging effect on twentieth-century thought and practice?" is posed, it seems that there are several good candidates—Nietzsche, perhaps, with his declaration of the death of God, or Marx, who famously characterized religion as the opium of the people. Evangelical Christians are perfectly sincere, however, when they answer that it is Darwin, chosen by the Anglican church to lie beneath the great, and, in his unorthodox way, devout Newton, who is the real culprit.

From the perspective of almost the entire community of natural scientists world-wide, this continued resistance to Darwin is absurd. Biologists confidently proclaim that Darwin's theory of evolution by natural selection is as well established as any of the major theories in contemporary science, as the atomic chemistry that schoolchildren learn or the molecular genetics that is emerging from a great scientific revolution still in process. Perhaps with a modest amount of overstatement, they echo Theodosius Dobzhansky's famous line, "Nothing in biology makes sense except in the light of evolution."[3] Religious scientists, often endorsing the ecumenical attitudes that accompanied Darwin's burial, express regret that their more militant fellow believers conjure an opposition that does not exist. Yet the issue will not go away. Detailed replies elaborated in one generation may inaugurate a period of calm, while resentment of Darwin and the establishment that defends him smolders sullenly. But the antipathy to

Darwinian evolution runs so deep that sooner or later the responses will be forgotten, ignored, or evaded, and the controversy will erupt anew.

Why is this? The question has two parts. First, how can the allegedly massive evidence in favor of Darwin's central claims be overlooked? How, if facts reflect what confident scientists say, is even the illusion of a serious debate created? Darwin's detractors cling to the belief that the "massive evidence" is overblown, and that the enthronement of Darwinism among the genuinely established sciences is the triumph of atheistic materialism. They believe that this atheistic materialism has cunningly co-opted religious scientists who don't even realize they have been tricked. Like all comprehensive scientific theories, evolutionary theory has unresolved questions that challenge biologists. In order to address these challenges and those of Darwinism's detractors, a clear presentation of the evidential situation, a delineation of the grounds on which Darwinism rests, of the problems it faces and to which its opponents point, and an appraisal of the merits of potentially opposing viewpoints must be provided.

The second part of the question concerns the source of the vehement opposition. Why is it that this particular piece of science provokes such passions, requires such continual scrutiny, demands such constant reenactment of old battles? Again, those who would disinter Darwin have a favorite explanation. The sepulcher in Westminster is a screen and the enthusiasm for Darwin's "reverence" a whitewash. From the militant evangelical perspective, foolish Anglican churchmen were caught up in popular enthusiasm, and signed on to "life without God." They thought, of course, that they were only eliminating God from any direct role in the long history of life on our planet, operating in the venerable tradition that saw the Creator's action as remote, as a wise institution of initial conditions from which the universe, and life within it, could unfold by well-designed natural processes. In fact, however, they were accepting "life without God" in a far more dangerous sense, blindly overlooking the subversive implications of this particular conception of life's history, the denial of all purpose, all providence, and all spirituality. The second issue, then, revolves around the implications of Darwinism. How does it affect our understanding of ourselves, our place in the universe, our religious beliefs and aspirations?

In what follows, I hope to address both issues.

* * *

I write at a time when opposition to Darwin has a new face. Intelligent design, it is claimed, is not a religious perspective at all, but a genuine scientific alternative to Darwinian orthodoxy, something that could be taught alongside evolutionary theory in the high-school biology curriculum without raising any anxieties about teaching religion, and that could even provide schoolchildren with an "exciting event" on their "educational journey."[4] Those who support this proposal, and who wish to see it enacted as law, can be divided, for my purposes, into two main groups. There are the architects of intelligent design theory, the "intelligent design-ers" as I shall call them, and the citizens whose support they enlist. In appraising the ideas and advertisements of the intelligent design-ers, I do not mean to criticize the sincere and worried people who rally to their cause. Only in the final chapter of this essay shall I consider the sources of their concern.

Advertising intelligent design as independent of religious doctrine is accurate in one important sense. To claim that some kinds of organisms are products of intelligent design does not logically entail any conclusion about the existence of a deity, let alone any specific articles of Christian faith. From a legal perspective, however, what matters is whether there might be genuinely nonreligious reasons for advancing a proposed law. If nobody would support the law except on the basis of religious beliefs, then, in the pertinent sense, the law cannot be independent of religion.[5] On this score, there are ample reasons for worrying about measures intended to introduce intelligent design into the biology curriculum.

In the first place, the style of argument that permeates claims of intelligent design traces back to William Paley's *Natural Theology*—required reading for Cambridge undergraduates when Darwin was a student and explicitly intended as a response to the "atheistic" arguments of David Hume's posthumous *Dialogues Concerning Natural Religion*.[6] Second, studies of explicitly Christian writings about Darwinism have shown that as the fortunes of "scientific creationism"—the favored alternative of the 1970s and the 1980s—have waned, references to "creation science" have given way to citations of "intelligent design" without other perturbations of the prose.[7] Third, as the recent trial in Dover, Pennsylvania made clear, the support for teaching intelligent design in the local high school came from religious people who felt the need to campaign for an alternative to Darwinism that accorded with their faith. Finally, in the wake of the rebuke administered by the voters in the local elections, who replaced the members of the Dover School Board who had agitated for the inclusion of intelligent design in the curriculum, Pat Robertson himself issued a warning that this apparent repudiation of God would undermine any appeal to the Deity should some catastrophe strike the community.

Although there are grounds for suspicion, I shall treat intelligent design as its leaders characterize it, as a hypothesis put forward to identify and account for certain natural phenomena. The sociological fact that the hypothesis is welcomed by a significant number of Christians, and by some religious people of other faiths, does not make it an intrinsically religious doctrine. A proposal about the natural world need have no specifically religious content to be more compatible with particular religious ideas than its equally naturalistic rivals. When Galileo made his case for the motion of the earth around the sun, and his opponents argued that the earth is at rest, the alternative hypotheses concerned natural processes; nevertheless some Catholics, committed to a literal interpretation of the biblical passage in which Joshua successfully commanded the sun to stand still,[8] believed that the earth-centered account was more sympathetic to the articles of their faith.

The core of intelligent design, understood as a rival to current ideas in biology and the earth sciences,[9] consists of two major claims. The negative thesis is that some aspects of life and its history cannot be understood in terms of natural selection, conceived as Darwinian orthodoxy supposes. The positive thesis is that these aspects of life must be understood as effects of an alternative causal agency, one that is properly characterized as "intelligent." (It is simplest to refer to this alternative agency as "Intelligence," so long as we don't engage in illegitimate personification—Intelligence is simply some causal power that deserves to be thought of as, in some sense, intelligent.) You could easily expand these two claims to a bigger package, adding the explicit identification that Intelligence is a creative deity, a providential creative deity, or even the God of the Christian scriptures. Outside the biology classroom, the expansion is permissible. Inside, it is not. The goal of the current movement to install intelligent design as an alternative to Darwinian evolution, is to reform the curriculum so that the two-part hypothesis is explained by the biology teacher, something that can be done without suspicion of religious indoctrination. It is, of course, a convenient fact that the local preacher can add elements of the larger package when he instructs the uncorrupted youth on Sundays.

Because the advocates of the intelligent design theory also insist on the further claim that the two-part hypothesis is genuinely scientific—the "status claim," as I shall call it—they invite a strategy for response. It would appear that their legal arguments could be undercut easily by rebutting the status claim, by showing that their favored two-part hypothesis isn't science. My own approach will proceed differently. I shall view intelligent design as "dead science," a doctrine that once had its day in scientific inquiry and discussion, but that has rightly been discarded.

It is easy to understand why many scientists (and the journalists to whom they give interviews) find the "not science" strategy attractive. After all, it is a quick way of dismissing the opposition, one that shortcuts the tedious work of analyzing the proliferating texts the opponents produce. But I think it can only succeed when the central issues are blurred. If the substance of the charge is that intelligent design is not science because it is religion, then the acquitting response should be, first, that the position can be formulated without making any religious claim (intelligent design is the two-part thesis just distinguished). Second, for much of the history of inquiry great scientists have advanced specifically religious hypotheses and theories. On the other hand, if we suppose that the two-part thesis doesn't have the characteristics required of "genuine science," then it is appropriate to ask just what these characteristics are. True, the architects of intelligent design don't spend a great deal of time performing experiments—but then neither do many astronomers, theoretical physicists, oceanographers, or students of animal behavior. Science has room for field observers, mathematical modelers, as well as experimentalists. Social criteria for genuine science, such as publishing articles in "peer-reviewed journals," are easy to mimic. Any group that aspires to the title can institute the pertinent procedures. Hence those procedures no longer function to distinguish science from everything else. So, what is left?

Many scientists believe that there is a magic formula, an incantation they can utter to dispel the claims of intelligent design. Indeed, intoning the mantra "science is testable," in the public press or even in the courtroom can produce striking effects. This, however, is only because of an overly simple understanding of testability. When the proponent of intelligent design points to some collection of natural phenomena, declaring that these could not be products of Darwinian natural selection but must instead be the effects of a rival causal agent, Intelligence, it isn't directly obvious how to test the hypothesis advanced. Unfortunately, that is the nature of the core hypotheses of many important scientific theories. The same could have been said for the hypothesis that chemical reactions involve the breaking and forming of bonds between molecules, or for the hypothesis that the genetic material is DNA (or, in the case of some viruses, RNA), or any number of sweeping assertions about things remote from everyday observation, when those hypotheses were first introduced. When such core hypotheses are tested, they are supplemented with other principles that explain the relationship between the core hypothesis and the processes to which we can gain observational access. To test the hypothesis that the genetic material is DNA, the pioneering molecular biologists of the 1940s needed a host of assumptions about just what was being transferred into the modified bacteria on which they performed their experiments. And it took several years' of ingenious laboratory work to show that those assumptions were justified. Thanks to their pioneering efforts, their successors were equipped with refined methods of detecting the observationally remote entities that figure in the hypotheses of molecular biology. More generally, the development of ways of detecting things that we cannot immediately see or handle is part of the creative work of science, work that expands our conception of what is observable. At many stages in the history of science, inquirers conceive of promising hypotheses that are hard to connect with observational or experimental findings. They and their successors must work to formulate auxiliary assumptions that will make the needed connections, assumptions that are often controversial, and that must be probed for their own soundness.[10]

Invocation of the magic formula thus faces a dilemma. If core hypotheses, taken in isolation, must be subjected to a requirement of testability to be taken seriously, then the greatest ideas in contemporary science will crumble along with intelligent design. If, on the other hand, all that is required is to supplement a core hypothesis with some auxiliary principles that allow for testing, then the spell fails to exorcise anything.

Unless it is shown that intelligent design, unlike the core principles of atomic chemistry and molecular genetics, cannot be equipped with auxiliary principles that allow it to be tested, then the charge of untestability will not stick. Moreover, demonstrating that intelligent design is not equipped with auxiliary principles requires detailed study of what the intelligent design-ers propose—that is, coming to terms with their confident positive claims that the operation of Intelligence can be detected in the history of life. In fact, when we do the detailed work of scrutinizing their claims, we shall discover why it is so tempting to dismiss them as "not doing science." It turns out to be difficult to connect the central theses of intelligent design with the observable evidence we have by deploying any principles that can be independently justified. But any right to dismissal cannot be assumed at the outset—instead, it must be earned.[11]

Fans of the mantra of testability will surely protest that my response to the friends of intelligent design is a cheat—and I sympathize with them. Simply crying "Foul!" however shouldn't convince a good referee. We must explain which rule of proper science has been broken, and how it has been broken. But pinpointing this explanation leads into thickets of philosophy from which no clear resolution has yet emerged. For the past half century, philosophers have tried and failed to produce a precise account of the distinction between science and pseudoscience. We cannot seem to articulate that essential line of demarcation.

There is, however, a deeper problem with the strategy of dismissing intelligent design as "not science." Intelligent design has deep roots in the history of cosmology, and of the earth and life sciences. Generations of brilliant and devout investigators firmly believed that their researches were supplements to the word of the Creator as revealed in sacred scripture, that they were disclosing that word by deciphering the Book of Nature. From Newton's speculations about the meaning of his "system of the world" to the country parsons who wrote about the fauna and flora in the parish precincts, there is a large body of work in "natural philosophy"—what we call "science," although the term was not then used in this sense—directed by the hypothesis of intelligent design, not in the modest two-part version, but in a theologically far richer package. If intelligent design is no longer science, it once was, and many scientific achievements we acknowledge build upon work that it inspired. Indeed, the status of intelligent design as a piece of mid-nineteenth-century science is confirmed by the many references in the *Origin* to "the theory of independent creation."[12]

Appreciation of the historical entanglement of science and religious doctrine should, I believe, incline us to the strategy I proposed above for responding to intelligent design. There is no place for intelligent design in the biology classroom because it is discarded science, dead science. From the perspective of reasonable educational policy, dead science only belongs in the curriculum in cases where a review of it is valuable for understanding live science. We study classical Newtonian mechanics because doing so is a necessary prelude to understand even the rudiments of quantum mechanics and relativity theory. The mere absence of a pedagogical need, however, provides no legal basis for exclusion. To show that intelligent design doesn't belong in the biology classroom would require arguing that the sole motivation for introducing this particular piece of dead science is a religious one.[13]

If some oddly motivated group were to campaign for teaching alchemy alongside modern chemistry, or the theory that heat is a "subtle fluid" in conjunction with thermodynamics, the right counter would not be to declare that these doctrines are intrinsically religious, or are pseudosciences. Instead we would explain that, although they once figured in science and were actively pursued by learned people, we have since discovered that they are incorrect, and that, if they belong in the curriculum, it should be in the history of science course, not in the chemistry or physics class.

Pursue the fantasy a little further, and imagine that the earnest activists disagree with the judgment that their pet theories are dead science. How would we try to persuade them? Surely we would do so by showing them the evidence that originally led to the rejection of alchemy and of "subtle fluid" ideas about heat. That wouldn't be enough, however. If our interlocutors were astute, they would remind us that hypotheses once abandoned can make a comeback. After all, Copernicus revived ancient views of the motion of the earth, and nineteenth-century physicists resurrected the doctrine, periodically fashionable since the dawn of science, that heat is the motion of the minute parts of matter. So we also would need to show how the further development of the sciences, after the activists' favorite ideas were given up, has reinforced that original judgment, how the evidence in favor of the orthodoxy that triumphed has continued to increase, how the issue looks from the perspective of the present.

If an appropriate response to advocates of discarded theories involves adjudicating an old debate from the vantage point of newer, even up-to-the-minute knowledge, can we manage without the history entirely? Would it be enough to ignore the considerations that initially led to discarding pre-Darwinian ideas, and simply explain how things now look? I think not. There should be no suspicion that the original decision was unreasonable, that it's just a fluke that things have gone well for a theory that gained an undeserved victory. Darwin's defenders don't suppose that previous attempts to reevaluate evolutionary ideas were wrongly dismissed, that it's only now, when scientific orthodoxy has a plausible tale to tell, that the orthodox can afford to come clean.

Hence, if we meet the challenge to Darwinian ideas, it will be necessary to understand how the central ideas of Darwinian evolutionary theory came to be accepted, and how they have fared in light of an increasing body of knowledge about the details of life on our planet. We need a historical perspective that leads us from the period during which the ideas espoused by the intelligent design-ers were widely accepted, through the episodes in which they were discarded in favor of Darwinian principles, to our present situation. We need, in short, to understand why intelligent design and other alternatives to Darwinism died, and why, despite the energetic efforts of the resurrection men, they have stayed dead.

* * *

Recertifying the demise of the allegedly live alternatives to Darwinism is more complicated than I have so far made it appear, because current opposition to Darwin involves three different debates. As we shall see, the most sophisticated of Darwin's detractors profit by intertwining them. "Darwinism" is not a monolithic whole, and one of the ideas that anti-Darwinians attack is by no means original with Darwin. If the biology curriculum is to be made thoroughly safe for Christianity, as the most vocal would-be reformers (but by no means all Christians) understand their religion, then there are three major principles that must be banned, or for which "balancing" rivals must be found.

The first of these principles is the idea of an ancient earth, a planet on which life has existed for almost four billion years and that has been populated at different periods by a large number of different kinds of organisms, the overwhelming majority of which are now extinct. Many of the animals and plants we know, including the birds and the flowers in our gardens, the wild counterparts of the living things we have domesticated, and our own species, are very recent arrivals in the history of life. Taken by itself, this first thesis leaves open the possibility that the history of life is one in which creation occurs in successive stages.

That possibility is explicitly denied by the second major principle, one that, unlike the first, was proposed and defended by Darwin. There is just one tree of life. All the living things that have ever existed on our planet are linked by processes of "descent with modification," so that even the organisms that seem least similar—an eagle and a seaweed, say—are derived from a common ancestor that lived at some point in the remote past.[14]

The last important idea, also central to Darwin's thought, concerns the causal processes that have given rise to the diversity of life. The principal agent of evolution, the chief cause of the modified descendants is natural selection. For any kind of organism, there will be variation in the descendants produced in each generation. Some of these variants will better enable organisms to survive the challenges of the environment, to mature, and to produce offspring. If the new characteristics that underlie their success are heritable, their descendants will enjoy the same good fortune, and the characteristics will spread. So, over a sequence of generations, a trait that was once rare may become prevalent.[15]

It's important to distinguish these three principles, because there are more or less ambitious ways of attacking "Darwinism." The most sweeping is to deny all three, to advance an alternative view according to which the earth is relatively young and has been populated, from the beginning, by the major kinds of plants and animals, including human beings, all created distinctly. Because this denial would allow for the narrative of Genesis to stand as the literal truth about the history of life, I shall refer to the position as "Genesis creationism."

A more modest conception, one that concedes that parts of the Bible's first book need not be read literally, would accept the ancient age of the earth but challenge the relatedness of all living things and the power of natural selection, at least in the most important events in the history of life. This kind of opposition to Darwin might well allow that plenty of organisms are related to one another by descent with modification, and that natural selection does sometimes, even normally, play a role in such processes. But it would insist on moments in the history of life where something else, something distinctly different happens, where new forms are created. In effect, the opponent supposes that there are breaks in the tree of life, alleged evolutionary transitions involving creative activity that generates something entirely novel—perhaps, for example, when multicellular organisms are produced, or when land-dwelling animals emerge, or when human beings originate. Since the fundamental idea is that the major novelties in the history of life are the products of creation, I'll call this approach "novelty creationism."

Contemporary versions of Darwinism conceive of life as having a single origin, from which living things split into distinct forms, called species, in events of speciation. These are the moments where the tree of life branches, sometimes identified when naturalists perceive differences they take to be significant, sometimes viewed in terms of interruptions of free interbreeding among the descendants of organisms who had previously mated quite happily with one another.[16] Novelty creationists today typically allow that there are (many) cases in which a species splits into two new ones, confining their attention to those changes that strike them as really significant.

Darwin thought in terms of a graceful tree of life, with relatively few branches. His modern descendants conceive of something bushier, a dense tree with large numbers of stubby branches representing dead ends, life's many failures. At a minimum, Novelty Creationism envisages the "broken tree of life," in which the gaps are bridged by some new act of creation (or, to speak unofficially, by the hand of the Creator). This vision can easily glide into that of a "garden of life," a scenario in which, while the earth is old, there are many separate acts of creation, many different, variously broken, trees. Or it can even become that of the "shrubbery of life," where, not so very long ago, a number of separate plantings were made at much the same time—a view that can endorse the narrative accuracy of Genesis.

Finally, the least ambitious of the challenges to "Darwinism" adopts both the thesis of an ancient earth and the thesis of the relatedness of all living things, the single bushy tree of life, while denying that natural selection has the power to bring about the major transitions in the history of life. Proponents of this idea point to the same episodes in life's history that serve their novelty creationist brethren as points of departure, the episodes in which something genuinely new seems to happen, something so complex that it couldn't, so the story goes, be the product of a blind and clumsy process like natural selection. Unlike novelty creationists, they allow that the complex forms that emerge are descendants of significantly less complex ancestors, denying only that natural selection could have been responsible for the change. In a sense, there is still room for something like "creative activity" but the products of that activity are new traits, organs, or structures in the descendants of ancestors who lacked such characteristics, rather than newly created whole organisms. This is the core of the official position of leading champions of intelligent design, and I shall call it "anti-selectionism."[17]

Marching under the banner of anti-selectionism gives one an air of respectability, because anti-selectionism has been vigorously championed by prominent evolutionary biologists in the past and is explored by some contemporary scientists whose (nontraditional) proposals engage the serious attention of their theoretically inclined colleagues. To wonder if a proposed cause is adequate to produce particular effects shouldn't earn excommunication from the scientific community. Indeed, if Darwin's detractors were merely to ask for some brief classroom discussion of currently unsolved problems in applying natural selection to the history of life, or even a simplified presentation of some alternative ideas about the origins of natural variations, thoughtful scientists and educators might welcome the suggestion. In general, and not simply in the case of evolutionary theory, it might be sound educational policy to identify places where there is further scientific work to be done. That is very different from taking seriously the thought that currently unsolved problems are doomed to remain unsolvable, and that there is a serious possibility that the entire framework of Darwinism should be discarded. The obvious and uncontroversial ways of presenting alternatives, or supplements, to natural selection would not do what Darwin's opponents want though, for they would make no mention of either intelligence or design.

So, while anti-selectionism might be central to the intelligent design movement,[18] at least in its conversations with the hitherto unconverted, it doesn't offer much to the faithful, for whom Darwin is still the bogeyman. What is necessary is a distinctive way of addressing those evolutionary transitions that so far nobody has explained by appealing to natural selection—a causal agency at work that genuinely deserves the label "Intelligence." This agency bestows on descendants traits, organs, and structures that were lacking in their ancestors. Merely applying some other natural process that complements, or substitutes for, natural selection in the problematic instances won't yield a rival vision that will be friendlier to faith than the current Darwinian orthodoxy. What many troubled Christians would like is some indication of planning, purpose, design, at work in the history of life, a providential hand that reaches in and produces the truly important changes.

As the case for intelligent design is elaborated, therefore, the position slides away from bare anti-selectionism toward the religiously more evocative position of novelty creationism. Instead of simply supposing that the great transitions in evolution—like the conquest of land, or the arrival of *Homo sapiens*—require something more than (or different from) Darwinian natural selection, there's a tendency not to see these as evolutionary transitions at all, but as episodes in which a genuinely creative Intelligence is active. The label, "intelligent design," is a brilliant cover for the oscillating target that so frustrates the scientists who rise to Darwin's defense, inspiring them to charge that intelligent design is not science. Although the label can stand for those special moments where the Creator's hand reaches in, it can also be divested of religious content, explained as merely a commitment to anti-selectionism.

In differentiating various positions for Darwin's detractors, I aim to bring clarity to a debate too often confused by their oscillations. There are three types of positions to be considered: first, anti-selectionism, that only opposes the sufficiency of natural selection to produce the major transitions in the history of life; second, novelty creationism, that takes some alleged transitions to be episodes in which organisms with new complex forms are created; and third, Genesis creationism, that hopes to make biology and geology safe for the literal truth of the Genesis narrative. Intelligent design presses toward novelty creationism when it can, retreating to anti-selectionism when the accusations of mixing religion with science roll in.

For many of those who want an alternative to Darwin, however, novelty creationism is not enough. They would remain shocked by a science curriculum that implied that any (nonpoetic) part of the Bible cannot be taken as literal truth.[19] If they clearly understood what the intelligent design movement would achieve, were it successful, these people would only be partly satisfied. Nevertheless, they might welcome the erosion of Darwinism in hopes that it could eventually lead to the triumph of Genesis creationism.

At different stages in the history of inquiry, each of the three positions had its day as part of scientific orthodoxy. Eighteenth-century discussions of the earth and its history typically took it for granted that the natural processes that had occurred would conform to the history related in the early chapters of Genesis. Only at the end of the century were there nascent suspicions that biblical chronology might be radically mistaken. In the first decades of the nineteenth century, however, those doubts multiplied, and by the early 1830s the claim that human beings had been present on earth ever since the dawn of life had become indefensible.

For the next decades, something like novelty creationism held sway, as the prevalent view maintained that the history of life on earth was a sequence of periods in which new life forms were first specially created, flourished for a while, and then went extinct. The culmination of the sequence was the most recent creation in which our own species was generated. In 1859, the publication of the *Origin of Species* began the vigorous debate in which novelty creationism was overthrown, and by the early 1870s, in the English-speaking world as well as in Germany and Russia, most researchers had accepted Darwin's conception of a single tree of life in which organisms are linked by processes of descent with modification.

Natural selection, however, remained controversial. Without a developed account of the mechanism of inheritance, it was quite unclear whether selection could give rise to significantly modified descendants. There were worries that the timescale for the history of life was too short for the observed diversity to evolve under natural selection, as well as perplexities about the power of selection to produce various types of traits and structures. Eventually, in the 1920s and the 1930s, biologists would produce the "modern synthesis," integrating Darwin's ideas about selection with the new genetics descending from Mendel's neglected work. Between 1870 and 1920, however, anti-selectionism was widely accepted, as biologists struggled to identify alternative mechanisms that would propel evolutionary change.

Around 1830, 1870, and 1930, respectively, Genesis creationism, novelty creationism, and anti-selectionism were discarded, consigned to the large vault of dead science. Are any of them ripe for resurrection? No. But the efforts of the resurrection men demand a rewrite of the obituaries, one that will expose clearly what happened and why the original verdicts have been sustained by the subsequent course of inquiry, or, in the case of anti-selectionism, why current worries about the power of natural selection hold no comfort for ideas of intelligent design. I shall try to show how contemporary evolutionary biology has come to its prevailing orthodoxies, and why there are no reasons to amend them in ways that would be welcomed by those who wish to disinter Darwin.

* * *

Yet the zeal of the resurrection men, and of the citizens who support them, also needs explanation. What threat, what real danger, does the acceptance of Darwinism pose? Even after a review of the evidence has shown that we are stuck with Darwin, we still need to decide whether or not his "profoundly reverent spirit" leaves central religious doctrines and cherished beliefs about ourselves unperturbed—whether, in short, that memorial in the Abbey undermines the institution and the values the site represents, whether, in the interests of accurate representation, Darwin really should be disinterred.

In the closing chapter of this essay, I shall argue that the thoughtful and concerned people who welcome the proposals of the intelligent design-ers have seen something important, that they fear, quite understandably, that they cannot live with Darwin. I shall try to show how evolutionary ideas combine with other bodies of knowledge to yield serious consequences for the future of faith—and how also a brusque strategy of dismissing superstition cannot be adequate. In the end, even after we have seen all the failures of the resurrection men, we cannot be content either with the well-intentioned accommodations of Darwin's Anglican eulogists or with the militant campaign to replace religion by erecting statues of the Sage of Down in every public square.

Chapter Two

GOODBYE TO GENESIS

Like several other texts from the ancient Near East, the Bible recalls a great flood, in which virtually all of the living creatures of the earth were destroyed. In chapters 7 and 8 of Genesis, we are told that it rained for 40 days and 40 nights, that the fountains of the deep were opened, that the waters covered the tops of the mountains, that every living thing upon the face of the ground, "man and animals and creeping things," perished in the deluge, and that the only survivors were the inhabitants of the ark that Noah had built and stocked. After 150 days the waters subsided, and, some months later, when the summit of Mount Ararat was uncovered, the ark finally came to rest upon it. Eventually the land became dry again, and Noah, his family, and his company of animals were able to emerge from their refuge.

Although the seventeenth-century divine, Thomas Burnet, believed that this story was literally and strictly true, in all its details, he discerned in it puzzles that called for scientific explanation. Prominent among them was the issue of the drying-up of the earth. Where did all the water go? How exactly was "the pond drained"? Burnet set himself the task of providing a convincing mechanical explanation, and, in 1681, he presented his conclusions in a learned and often ingenious book, *The Sacred Theory of the Earth*.

It is easy to smile indulgently at the earnest struggle to show the possibility of processes that Burnet believes must have occurred because sacred scripture reveals to him that they once took place, the detail lavished on the resulting difficulties, the intricate diagrams and calculations. Yet it is a serious research project, one undertaken with integrity and honesty, and one fully representative of the early modern scientific temper. Most of the great figures who made the principal discoveries of the episode, or series of episodes, known as the "scientific revolution"—Copernicus, Kepler, Boyle, and Newton, to cite just four examples—firmly believed that one important point of their work was to display the wisdom of the Creator, to "think God's thoughts after Him." They, like Burnet, believed in two routes to God, one through the scriptures, the revealed word, divinely inspired, and one through the Book of Nature, in which the discerning eye would see, and understand, divine providence at work.

Although commentators from antiquity on had adopted nonliteral approaches to the scriptures, approaching sacred texts as sources of spiritual truths, many seventeenth-century thinkers, Burnet and Newton among them, were committed to literal interpretations. Perhaps the long, and bloody, conflicts between Protestants and Catholics, centered on different interpretations of the same Bible, forced a new concern with principles for reading the scriptures. In any event, for the first time in the history of Christianity, an emphasis on literal interpretation became dominant, giving rise to the kind of problem that troubled Burnet. Truth cannot be in conflict with itself, so that the two books—the Book of Nature and the Word of God in the Bible—must be compatible with one another, even when the scriptures are read literally. It is thus a proper project for the reverent investigator to address those perplexing instances in which reconciliation appears difficult—to explain, for example, how the pond was drained.

For more than a century after Burnet wrote, earth science and natural history would be dominated by the felt need for reconciliation with scripture. Country clergymen explored the landscape in the vicinity of their churches, describing the plants and animals they found, and offering their reflections on the subtle ways in which the Creator had adapted them to their way of life. So, for example, Gilbert White wrote in *A Natural History of Selbourne*, "To a thinking mind nothing is more wonderful than that early instinct which impresses young animals with the notion of the situation of their natural weapons, and of using them properly in their own defence, even before those weapons subsist or are formed. Thus a young cock will spar at his adversary before his spurs are grown; and a calf or lamb will push with their heads before their horns are sprouted."[20] The "thinking mind" appreciates the wisdom of God in the creation.

Harmony between the Bible, literally understood, and the embryonic studies of the earth and its inhabitants imposed constraints on those investigations. Some biblical scholars used the genealogies and chronologies of the Old Testament to calculate an age for the earth, usually assigned as six thousand or so years. Although it was difficult for eighteenth-century thinkers to envisage how these estimates might come into conflict with evidence from the field, potential disagreement might be resolved by scrutinizing the specific assumptions made in the calculations—after all, scripture does not explicitly declare an age for our planet. Two claims that flow more directly from the early chapters of Genesis are more difficult to evade. First, Genesis states that all major kinds of plants and animals, as well as human beings, were created at the beginning, and that all have lived on the earth continuously throughout its entire history. Second, the Bible says that there was once a great flood in which almost all living things were destroyed, and that all the organisms that have lived since are descendants of the small company that survived the flood. One of the surprises of eighteenth- and early nineteenth-century investigations proved to be that these claims appeared incompatible with the most obvious explanation of the rock strata.

As more and more curious structures bearing an uncanny resemblance to shells, teeth, and bones were discovered, it became apparent that fossils were residues of organisms that had lived in the past. Recognizing the division of rocks into layers, geologists came to accept the principle of superposition, interpreting deeper levels as older than more superficial strata. In the early nineteenth century, the hydrologist William Smith, who often served canal builders as a drainage consultant, began to correlate rock types across regions, relying first on criteria of chemical and structural similarity, and, for more refined connections, on the observed distinctness of the fossils associated with different strata. (It is important that Smith had a prior stratigraphical method to use to correlate the rocks, which he could deploy to demonstrate the principle that strata contain their own distinctive types of fossils. With that principle in place, he could then use these fossils as indices of identity and difference across a much wider range of strata.) Smith's pioneering work in geological mapping inspired others to try to show how the rock layers in different regions related to one another—to produce, in the end, a scheme connecting and ordering the strata across the entire globe. Even in the relatively early stages in the 1820s, however, it was apparent that there was a consistent sequence of fossil types, from the deepest (oldest) to the most superficial (youngest), and the appreciation of the sequence prompted a revision of the biblical narrative of the history of life.

Here is the obvious story with which the devout geologists of the early nineteenth century were confronted. Sedimentation rates suggest that the age of the earth is much greater than hitherto supposed, for those rates would require vast stretches of time (at least hundreds of thousands of years) to lay down rocks to the depth observed. These strata were deposited sequentially, and the oldest almost always lie at the bottom. Most of the organisms they contain belong to species that have now vanished from the earth. The deepest rocks (more exactly, the deepest fossil-bearing rocks then known) contain residues of marine invertebrates, some of which, like mollusks, are very familiar, others of which, like the trilobites, are very different. Above them lie layers in which there are both marine invertebrates and some fish, with an increasing diversity of fish as you climb the rock column. Higher still are strata with marine invertebrates, fish, and amphibians. Ascending further, these kinds of organisms are joined by reptiles, including huge reptiles of kinds that no longer exist—some of the dinosaurs—and later, after the vanishing of the dinosaurs, by birds and mammals, both of which become increasingly diverse as you approach the top. Near the surface, in the shallows of the rocks, there are finally traces of apes, and, eventually, of human beings.

The plant record displays a similar, uniform, pattern. The earliest strata that bear plant remains contain residues of ferns, and, at higher levels, they are joined first by conifers, and later by deciduous trees and flowering plants. The picture, in both instances, is of a sequential history of life, one that belies the idea that all the major kinds of plants and animals have lived on the earth since the very beginning.

So it is hardly surprising that Adam Sedgwick, the Reverend Adam Sedgwick, a man who would guide the geological apprenticeship of the young Charles Darwin, who would write a monstrously long (400-page), fiercely negative, review of a pre-Darwinian evolutionary treatise, who would be highly critical of Darwin's *Origin*, who had fought hard against a history of life at odds with Genesis, eventually had to give up. In 1831, he delivered his presidential address to the Geological Society:

> Having been myself a believer, and, to the best of my power, a propagator of what I now regard as a philosophic heresy . . . I think it right, as one of my last acts before I quit this Chair, thus publicly to read my recantation.
>
> We ought, indeed, to have paused before we first adopted the diluvian theory, and referred all our old superficial gravel to the Mosaic Flood. For of man, and the works of his hands, we have not yet found a single trace among the remnants of a former world entombed in these ancient deposits. In classing together distant unknown formations under one name; in giving them a simultaneous origin, and in determining their date, not by the

> organic remains we had discovered, but by those we expected hypothetically hereafter to discover, in them; we have given one more example of the passion with which the mind fastens upon general conclusions, and of the readiness with which it leaves the consideration of unconnected truths.[21]

It was a funeral oration for the kind of geology that Burnet and his successors had practiced, a geology dedicated to harmony between nature and scripture.

* * *

The problem is obvious. How do you explain this consistent ordering of fossil remains if not as what it seems to be, namely a sequence of episodes in the history of life showing very different organisms at different stages? How do you account for the fact that the remains of the kinds of organisms that now exist, the birds and trees and flowers and mammals we know—not to mention human beings—are found only in the most recent deposits? The Book of Nature appears forthright, and forthrightly incompatible with Genesis.

Sedgwick's reference to the "diluvian theory" and the "Mosaic Flood" points to the line of response he had tried, and that he had to "recant." Even though Genesis begins with the roughly simultaneous creation of the flora and fauna of Eden, it does contain a story about a later event that might have played havoc with the rock record. If almost all earth's life drowned in a roiling flood, the ordering of the fossils might be an effect of the deluge. Perhaps we should return to—and extend—Burnet's mechanics of the flood.

Because the ordering of the strata is uniform across widely separated regions, it is important that there be some general principles that account for the distribution of the fossils. It won't be enough simply to declare that the waters produced a chaotic mix. Instead, we need to understand why the birds are always found at the top and the fish appear originally near the bottom.

At first sight, there seems to be an obvious answer. The fish live in water and were entombed early in the masses of loose sediment produced by the opening of the "fountains of the deep." The birds, on the other hand, could escape by flying away, perching to rest on whatever pieces of land remained, until even the mountaintops were covered. Indeed, fans of "flood geology," from the early nineteenth century to the "scientific creationists" of our own day, have suggested that understanding the environments in which plants and animals lived, and thus where they were when the flood waters hit, will make sense of the uniform pattern found in the rock column. Troubles immediately arise, however, when you descend from the general idea and consider the details.

Specimens of the class that includes most contemporary fish are found in the fossil record from rocks whose age is estimated to be about 200 million years and upward into the present. Much deeper deposits contain the remains of types of fish that no longer exist, as well as fossils of sharks. By contrast, fossil whales, dolphins, seals, and porpoises are found only near the top of the rock column. Among the contemporary types of fish, preserved in the same strata, are species that inhabit the depths of the ocean, species that live near the ocean surface, species that live in lakes and rivers. So we can rule out the idea of a fine-grained sorting by location. Nor can it be supposed that the fish buried in the higher strata must have been especially lithe, and thus able to evade the sediments that hurtled through the waters, at least for a time. Not only are some of the remains found in recent deposits fossils of bottom-dwelling fish, but the fish in question are large and sedentary. Moreover, the contrast between sharks on the one hand, and dolphins on the other, is quite perplexing. Sharks and dolphins occupy similar environments, and have roughly equivalent capacities for moving speedily through the water. Yet plenty of sharks, but no dolphins, were buried at relatively early stages of the flood.

The puzzles increase when we consider the birds. Some birds—ostriches and penguins, for example—are flightless, and would appear to be strong candidates for an early entombment. Others live in marshy areas that would have been inundated quite early, and among these are kinds that lack the capacity to fly far. As the floodwaters rose, there would apparently have been increasing competition for resting places, and we might expect larger birds, with greater capacity for staying aloft and for defeating rivals, to be the last ones to go. The final potential perch would have been the ark itself, and we can only imagine the Hitchcockian scene in which Noah and his company found themselves.

The fossil record of the birds shows none of the order flood geology leads us to expect, and, when we turn to the plants, there is further discomfort. Why do we find ferns in relatively ancient strata? Why are they joined first by the conifers and only later by deciduous trees and flowering plants? To take the principle of assortment by upward mobility seriously conjures a vision of the plants struggling uphill away from the rising floodwaters. An oak tree leads the way, as a pine labors behind, and an overweight fern puffs in the rear.[22]

As Sedgwick reluctantly concluded, the detailed evidence from the rocks dooms the idea of explaining the distribution of fossils by appeal to a universal flood. The regularities in the sequence, of which I have mentioned only a few of the most striking types, call out for explanation. Why are these kinds of fossils always found at deeper levels than those? Appreciation of the regularities, both in a gross form (ferns before conifers before deciduous trees) and in more refined ways (small three-toed horselike animals before horses), only exacerbates the difficulties for saving Genesis creationism.

In fact, the invocation of Noah introduces further problems. Some living things were preserved by carrying them on the ark. How many, exactly, and which ones? If representatives of all modern species were included, the task of assembling them, leading them to their proper places on the ark, and keeping them alive during the months afloat would have been immense. Predators and prey would have to be carefully separated. Very different environments would have to be provided for polar bears and camels. Noah and his seven human helpers would have had to be in constant motion from stall to stall, providing food and undertaking the Herculean task of cleansing the stables.[23]

When you consider the occupants of the ark, it's natural to focus on the animals most salient to us, the ones the Genesis narrative explicitly mentions. You forget that at least some plants and fungi would have to be carried along, too, that cacti and orchids, willows and mushrooms, would have to have appropriate environments and proper care. Even less salient are numerous microorganisms, including bacteria, viruses, and other parasites, many of which are specific to particular plants, or animals, or humans. So, as we imagine Noah and his family pursuing their whirling round of chores, we have to suppose that they are harboring all sorts of infectious agents, that one of them has cholera, one diphtheria, several various respiratory infections, and so on—and that either one of the people, or maybe some dog or squirrel or bat among the company, is rabid.

Once they arrive on Mount Ararat, can they finally rest? Not really. For, if they are to produce living creatures that will repopulate the earth, then they must be extremely careful to maintain the separation of predator from prey. It won't do for the intended grandmother of the future gazelles to disappear into the mouth of one of the ancestral cheetahs. The animals will have to be led carefully to points from which they can reach their intended destinations—so that the marsupials can make their way to Australia in the next five thousand years or so (probably a forced march for the more sedentary ones like koalas and wombats), the polar bears to the Arctic, the llamas to South America. The plants will have to be treated carefully to ensure that they reach regions in which they can grow and thrive. Apparently, the ark's crew of eight—or less, if cholera, diphtheria, and other diseases have done their deadly work—will be in for significant traveling.

There is an obvious way to decrease the workload for this overburdened band, to cleave more closely to the inventory provided in Genesis, supposing that the ark only contained representatives of particular "kinds," and that, after the flood waters receded, there was a process of diversification within "kinds." Noah didn't need to line up pairs of all the species—it was enough to take pairs of the "dog kind," the "cat kind," the "deer kind," the "snake kind," and so forth. After the flood, the dog kind has diversified to give rise to wolves, coyotes, jackals, and foxes. The representative of the snake kind has been the source of hundreds of species of snakes.

To the extent that this idea cuts down on the necessary labor, it faces an obvious objection. Genesis creationism supposes that about five thousand years elapsed between Noah and the present. The process of diversification must thus be extremely rapid. Suppose it occurs through the splitting of kinds, and the descendant groups to which they give rise, so that every so often a group gives rise to two descendants, and that all the kinds and the groups they generate endure. So, in eight events of splitting, you could obtain just over five hundred species from an original "kind."[24] The average rate at which splitting occurs is thus about once every 625 years. Although Darwin's detractors complain that the diversity of life couldn't possibly be produced at the speed Darwinians suppose, they seem to envisage a rate of evolution (or "diversification within a kind") extraordinarily faster than any biologist has ever dreamed of—and to overlook the fact that none of this "diversification" has been recorded in human history.

The only way to resurrect Genesis creationism is to consider these problems soberly and seriously, and to account for the ocean of anomalies, as Sedgwick once tried to respond to the new fossil evidence, and Burnet attempted to explain the mechanics of the flood. Until the hard work has been done, there is no reason whatsoever to question the verdict of the 1820s. Devout geologists reluctantly said goodbye to Genesis. They were honest—and they were right.

* * *

Genesis creationism lurks in the background of the creationist movement, occasionally surfacing in the writings of the scientific creationists of the 1970s and 1980s. But it is not visible in the public face of intelligent design, the movement that now attracts the vast majority of those who once supported introducing "creation science" in their schools. There's a lingering tendency to think of the Genesis narrative and the standard geological account of the history of the earth as alternative hypotheses about the past, deserving "balanced treatment," a tendency made vivid by a double picture on the front page of the *New York Times*, depicting two distinct groups of visitors to the Grand Canyon, each of which offered radically different explanations of the natural wonders they saw.[25] If the problems cut as deep as I have suggested, how can the suggestion of genuine alternative views elicit any reaction other than laughter?

Genesis creationism thrives on ignoring demands to work out systematic explanations that would address the flood of problems it confronts, a strategy that, as we shall see, pervades the intelligent design movement. Instead of articulating positive doctrine, rewriting *The Sacred History of the Earth*, it seeks out phenomena for which no orthodox account has yet been given. Here, we are told, the allegedly uniform series of strata is inverted or displaced. There, we find evidence of human remains and dinosaur bones in the same deposits. Elsewhere, in the Grand Canyon perhaps, the layers are disturbed in a fashion that indicates some cataclysmic event.

So rather than focusing on the mountain of instances that standard geology explains and that challenge any account compatible with Genesis, we're invited to focus on a molehill of puzzles, as yet unsolved. The patient response, true to the integrity of Burnet, Sedgwick, and a host of geologists since, is to address the puzzles. Champions of orthodox geology would explain how, in some instances, the strata can be displaced, folded, or inverted. They would identify the mechanical forces at work and point to the traces of their action. They would examine the fossils that are meant to show that human beings and dinosaurs walked the earth together, and uncover the misidentifications that "scientific creationists" have made. They would show how standard theories of deposition and the formation of strata allow for occasional events in which there are large modifications of the earth's surface, and how the phenomena indicated provide no evidence for a universal flood. That response has been elaborated by a significant number of responsible, and responsive, scientists, who have selflessly taken time from what they (reasonably) view as more important research to address the anti-Darwinian charges.

Friends of Darwin should be grateful for these patient responses, but they should also protest the perceived need to answer whatever challenge creationists pose. Until Genesis creationists address the full range of the evidence of regular succession in the rock record, until they elaborate a detailed explanation as to how Noah's ark served as the conduit through which the originally created living things left descendants in the world today, they have no genuine rival scientific hypothesis with comparable explanatory power to set beside an orthodox account of the history of our planet. An analogy may help to reinforce this point. The ravages of AIDS are caused by prior infection with the HIV virus, or so most people who have an opinion on the subject believe. Potential challenges based on the existence of a few people who have harbored the virus for many years but who have not developed symptoms of AIDS might be met with attempts to understand what factors have inhibited the normal progression of the disease in these cases. Of course, future control of the disease could benefit from knowing what distinguishes these more fortunate individuals. Yet, even before we attain this knowledge, there is no serious reason to doubt the standard account. Questioning the connection between HIV and AIDS on so slender a basis would be absurd—and dangerous—given the vast body of positive evidence that medicine has amassed in the past quarter century. Such skepticism could only be remotely plausible if the medical evidence were somehow ignored, cast into shadow, so that only the handful of puzzling cases commanded our attention.

Genesis creationism can only survive by well-publicized advertisement of "anomalies for Darwinism," and by relying on public ignorance of the grounds on which nineteenth-century geologists abandoned it. Even this strategy, however, proves hard to sustain. For contemporary geology has access to windows into the past unknown in the early nineteenth century. The painstaking work of stratigraphical correlation, of charting the sequence of fossils in the rocks, has been supported from an entirely independent science: physics.

Nowadays schoolchildren routinely learn that dates can be assigned to rocks by employing principles about radioactivity. Radioactive elements decay, yielding daughter elements (or decay products). By measuring the decay products and solving a few equations we can estimate the age of the stratum in which the products were found. This step is performed many times, employing different radioactive processes (processes that involve the decay of different initial elements), and the dates we calculate agree with one another—they are considered *concordant* in the terms of the trade. Thus we acquire a basis not only for ordering the strata, but also for creating a scale that assigns numerical ages to rocks at different levels.[26]

Two obvious problems beset radiometric dating. First, how can we know the amount of an element initially present in a sample? Second, how can we tell that the process hasn't been contaminated through the introduction of pieces of daughter elements that did not come from radioactive decay? Geophysicists address these questions in a variety of ways. They look for decay products that would not have been present when the materials to be dated were originally formed. Or they compute initial amounts by comparing different decay processes, using findings about the diffusion of potential interlopers to check that the system has remained uncontaminated, and so forth. Working out the details of different dating methods enables them to recognize which kinds of materials can be dated in which ways, and, by assembling a very large number of independent techniques for assigning ages, they can rule out the concern that their estimates of initial frequencies or their claims about absence of contamination are wrong.

This work has been done so thoroughly and in such detail that contemporary geology can talk with considerable confidence about the ages of the strata, and the times at which different kinds of living things emerged. It vindicates the conclusions drawn in the 1830s, showing that over almost four billion years—far longer than any nineteenth-century scientist could have anticipated—our planet has been home to a succession of floras and faunas, that most have become extinct, and that the species most salient to us (and to the authors of Genesis) are very recent arrivals. However gallantly it tilts against the evidence of the rock column, Genesis creationism today has to meet the hard challenge posed by radiometric dating.[27]

Since the 1980s, Genesis creationism has retreated into the background, and I suspect that few architects of intelligent design are attracted by the prospect of a debate with geophysicists about the physics of radioactive decay. Instead, they concentrate their fire elsewhere, and I anticipate the response that my review of the demise of Genesis creationism merely defends an orthodoxy they are happy to allow. I hope that the response will be made openly and publicly, cried from the rooftops, because, if intelligent design accepts the orthodox story of the stages of life's history, if it gives up the literal truth of Genesis, then its supporters should know this. They should not think that this movement will provide them with an ideal curriculum, one that will be fully compatible with the Christian scriptures as they understand them—or even that it will pave the way for some such curriculum in more godly times to come. If my discussion bluntly hammers away at points the proponents of intelligent design willingly concede, then they must believe, as I do, that Genesis creationism is dead, and that dead it will remain. Here, the would-be resurrection men meet the limits of their power.

Earlier critics of Darwin wanted far more than this, demanding a brief for biblical literalism. For, they declared, if the dam against nonliteral readings is once breached, the floodwaters of "complete apostasy" will sweep in.[28] The only occasions for straying from literal interpretation are when the form is obviously poetic (as in the Psalms) or when the word of God is plainly adapted to the human point of view (as in the problematic passage Joshua 10: 12–14, where Joshua commands the sun to stand still). Here, Darwin's opponents counter a venerable tradition of scriptural interpretation, one that recognizes many different kinds of narratives and texts within the Bible. In some Christian denominations, that tradition endures and the demand for literal reading is viewed as unnecessarily strict, even as quite misguided. But it isn't hard to understand the literalists' point. Suppose you abandon the literal truth of the

Creation story (or, more properly, stories—for there are two distinct versions at the beginning of Genesis), and the literal truth of the flood narrative. How much of the rest of Genesis also becomes suspect? There are, after all, a host of curious matters to contend with: the immense ages of the early protagonists, the puzzle of the mother of Cain's children, the postmenopausal conception of Isaac, the remarkable behavior at Sodom where Lot offers his virgin daughters to an angry mob in order to defend the two beautiful young men (angels) who have just arrived, the wrestling match between Jacob and the angel. If we know that certain parts of the book are not literally true, we might worry that these peculiar stories can't be taken at face value. Perhaps the allegorical parts of the Bible even extend beyond Genesis, so that we might wonder about Pharaoh's bizarre changes of mood as his nation is afflicted with successive plagues, about Elijah's miraculous confutation of the priests of Baal, about the virgin birth, about the vision on the road to Damascus, even about the resurrection itself. Devout people fear that, when thought through, traditions of nonliteral reading that have dominated much of the history of Christianity, and that are present in liberal Christianity today, will not leave much standing. How should we sort out what to take literally and what to take as allegory? Someone who poses the question may be led to investigate scholarly studies of the text, to learn the prevalent views about how it was composed and how this particular group of books became canonical—and, as we shall see, these investigations may prove profoundly disturbing.

In saying goodbye to Genesis, do we bid farewell to a lot more? I shall postpone addressing the question until the final chapter of this essay.

Chapter Three

ONE TREE OF LIFE

On the advice of his Cambridge mentor, John Henslow, Darwin chose the recently published first volume of Charles Lyell's *Principles of Geology*, a magisterial development of understanding the earth's past in terms of forces observable in the present, an account that summarized and extended the new orthodoxy that had replaced Genesis, as one of the few books to take on the voyage of the *Beagle*. Later, when the *Beagle* docked for the second time in Montevideo, Darwin received a copy of Lyell's second volume, devoted to the issue of whether the succession of organisms whose remains were found in the rocks testified to some process of "transmutation," or whether it represented episodes of successive creation and extinction. Lyell argued against the relatedness of living things, opting instead for the idea of periodic creation, a process he candidly described as the "mystery of mysteries."

Lyell saw the alternatives with impeccable clarity. If the earth's history reveals different floras and faunas at different stages, then there can only be two possibilities: either the later ones are modified descendants of the earlier ones, or they are somehow created anew. How could the issue be resolved? In retrospect, Lyell's ingenious arguments seem colored by his hopes at arriving at a psychologically and theologically comfortable perspective on the history of life. Given the apparent difficulty of deciding the issue either way, it is hardly surprising that Lyell's conclusions took the form they did—or, that a decade later Darwin would cautiously present his own "transmutationist" views to his close friend, Joseph Dalton Hooker, by declaring that it was like "confessing a murder."

The observations on the *Beagle* voyage, and his subsequent reflections on what he had seen, were partially responsible for Darwin's acceptance of a heretical view. Patiently, over the decades, he constructed a multifaceted case—"one long argument," as he called it—for a conclusion opposite to that reached by Lyell. The past history of life has left clues in the present, phenomena that can only be explained by supposing that living things are related to one another.

Although the *Origin* begins with a discussion of natural selection, and although Darwin needed a mechanism through which species might be modified, much of the argument of the book is devoted to comparing the two alternatives Lyell had delineated. The "long argument" depends on reviewing numerous details about living things, as they appear in the present and as the fossil record shows them to have been in the past. Darwin traces the similarities and differences among organisms, considers the places and environments in which they live, describes the variations that occur in particular groups, and so on. In each case he asks whether the details he reports make sense on the supposition of creation, or whether they are explicable only in terms of a history of descent from previous organisms.

The following passage is typical of hundreds that recur throughout the *Origin*:

> It is difficult to imagine conditions of life more similar than deep limestone caverns; so that on the common view of the blind animals having been separately created for the American and European caverns, close similarity in their organization and affinities might have been expected; but, as Schiödte and others have remarked, this is not the case, and the cave-insects of the two continents are not more closely allied than might have been expected from the general resemblance of the other inhabitants of North America and Europe. On my view we must suppose that American animals, having ordinary powers of vision, slowly migrated by successive generations from the outer world into the deeper and deeper recesses of the Kentucky caves, as did European animals into the caves of Europe.[29]

There are cave insects and ordinary, seeing, insects in North America, and also cave insects and ordinary, seeing, insects in Europe, and they occur in very similar physical environments. All these insects share some general features. What interests Darwin is the fact that, beyond these common traits, the American cave insects show more similarities with the American seeing insects than they do with the European cave insects, just as the European cave insects are more like the European seeing insects than they are like the American cave insects. Why is this? As Lyell saw, there are only two possible accounts of the origins of these groups. Either the cave insects were separately created or they are modified descendants of other organisms. Suppose that they were separately created. You might think that they would be created to fit their environments, that there would be some design—an intelligent design perhaps—that would constitute the "best plan" for a cave insect, or the best plan for the common physical environment, and that this would be found in cave insects across the world. That's not what the natural world reveals, however. Instead, the "design" of the North American cave insects looks like a modification of the design of North American insects generally, and, likewise, the design of the European cave insects looks like the design of other European insects. Whereas the proponent of special creation has to see this as a puzzling brute fact—an expression of the whimsy of creation—Darwin offers a simple explanation. The cave insects are modified[30] descendants of the insects who once lived in the vicinity, many of whose other descendants still buzz around in the open air outside the caverns, and that is why the American cave insects resemble the seeing American insects, and the European cave insects are like the seeing European insects.

I've reconstructed this particular argument in some detail to show how powerful it is. Given the rejection of Genesis creationism there are just two possible avenues for explaining the origins of groups of living things. Darwin shows that one of these, the "creationist avenue," leads nowhere, but that the alternative does account for a particular phenomenon—the apparent retention of the characteristics of other insects in the vicinity. Thus he undercuts the main creationist explanatory idea, to wit that the design of organisms is planned to fit their environments. So what? Creationists can surely find comfort in Lyell's resonant phrase—this is a "mystery of mysteries." They can reasonably declare, as far as the cave insects alone are concerned, that this is a curiosity. Who should care about these particular creatures? The argumentative weight of the *Origin*, the heft that crushed creationist objections during the 1860s, lies in the fact that this is not an isolated case. Again and again, Darwin offers arguments with a similar structure.

Consider the horses. As breeders know well, foals sometimes display coat patterns, bars or stripes, that resemble those found in zebras. Darwin reviews a range of examples, explaining them in terms of the descent of horses, asses, zebras, and other equine species from a common ancestor. He contrasts his account with that available on the alternative approach. "He who believes that each equine species was independently created, will, I presume, assert that each species has been created with a tendency to vary, both under nature and under domestication, in this particular manner, so as often to become striped like other species of the genus . . . To admit this view is, as it seems to me, to reject a real for an unreal, or at least for an unknown, cause. It makes the works of God a mere mockery and deception; I would almost as soon believe with the old and ignorant cosmogonists, that fossil shells had never lived, but had been created in stone so as to mock the shells now living on the sea-shore."[31] As with the cave insects, the creationist alternative is reduced to seeing, in the apparent retention of traits of other species, a whimsical creation, or, more piously, to avowing a "mystery of mysteries."

Why are there birds with webbed feet that live on dry land? Woodpeckers where no tree grows? Why are the fossils of extinct mammal species in Australia similar to the marsupials that inhabit the continent today? Why are the extinct armored mammals of South America akin to the currently living armadillos? Why are the birds of South America so like one another and so different from the birds of the Old World? Why does the same apply in the case of reptiles and mammals? Why do the floras and faunas of islands regularly resemble those of the neighboring continents? The examples mount. "What can be more curious than that the hand of a man, formed for grasping, that of a mole for digging, the leg of the horse, the paddle of the porpoise, and the wing of the bat, should all be constructed on the same pattern, and should include the same bones, in the same relative positions?"[32] Cases like this—and Darwin cites others concerning the relations of bones in the skull to vertebrae, jaw structure and leg structure in crustaceans, the inner structures of flowers—move him to a strong statement of the explanatory vacuity of creationist orthodoxy. As he points out, when challenged to explain these similarities by pointing to "utility or by the doctrine of final causes," staunch creationists have to admit that it is hopeless. They can only say that this is the way in which creation has occurred, "that it has so pleased the Creator to construct each animal and plant."[33]

That line becomes less plausible the more times the creationist is forced to utter it. For the fundamental point of Darwin's deluge of examples is that, in a vast range of instances, when you look closely at the "design" of plants and animals and assume that the designer started from scratch, it is not at all intelligent. If you were designing a porpoise paddle, a horse leg, a human hand, a mole forelimb, and a bat's wing, without any prior constraints, you could do a lot better by deviating from the common plan. To understand living things as manifesting intelligent design you have to appreciate the ingenuity with which the available materials have been modified to new purposes. It isn't simply that these designs are imperfect, but that a truly intelligent designer, liberated from any constraint to produce descendants from previous organisms, would be expected to do much better. Common descent shows up especially clearly in the many shared features of organisms that developed in quite different ways to meet the challenges of their environments.

Imagine that you were given the task of painting a series of pictures that would realistically depict a rapidly changing scene. Each day, what you need to represent is different from what confronted you the day before, and the differences increase with time. Each day you are given a fresh canvas, and, at the end of your project, you have a series of pictures.

You have a counterpart, someone who has a more difficult assignment. This person is only given one canvas, and must constantly paint and repaint. Some of the lines, those painted more recently, are relatively easy to amend or cover. Others, the older ones, resist concealment, becoming ever more indelible, and have to be used as elements of later pictures. At the end of each day, a Xerox copy is made of the picture produced.

Both of you finish, and the sequences of representations are compared both with one another, and with color photographs of the scenes you have been asked to depict. The pictures you have produced are quite different from the copies generated from your counterpart's work—yours are judged to be far more faithful representations of the scenes you were asked to paint. Those who believe you both had a completely free hand (unfairly) take your counterpart to be a bungler.

Central to the *Origin* is the idea that when you look at the details, the structures, situations, and relationships among living things are like your counterpart's pictures. To suppose that these living things were separately created is to view the creative agent as whimsical, bungling, a mediocre engineer, an unintelligent designer.

* * *

By 1867, Darwin could write to his friend and champion, Thomas Henry Huxley, expressing his satisfaction that the "main point," the relatedness of living things through descent with modification, had been widely accepted. Four years later, in an anonymous review, Huxley confidently judged that Darwin's *Origin* had worked as complete a revolution in natural history as Newton's *Principia* once achieved in astronomy. Just as, in the 1820s, many scholars had reluctantly given up on Genesis creationism, so too, in the 1860s, their counterparts, including Lyell himself, were forced to admit that Darwin had amassed far too many phenomena that could only be explained by supposing the relatedness of living things.

Yet there was much that Darwin had not been able to explain, and he candidly drew attention to some principal problems he could not solve. Particularly serious were questions to which later creationists—the scientific creationists of the 1970s and 1980s—and champions of intelligent design would point. Why does the fossil record not reveal the "numberless transitional links" presupposed by Darwin's explanations? What accounts for the apparently sudden explosion of life at the beginning of the Cambrian period? (Fossil remains of multicellular organisms are first found, in considerable profusion, in strata about 540 million years old.) How can Darwinians account for the origin of complex organs and structures—the eye being a favorite example?[34] Darwin's contemporaries knew of these difficulties, and raised them in reviews and debates about his theory. But they recognized the difference between the two Lyellian alternatives: one that accounted for nothing, the other that already explained a great deal and that held out, as Darwin saw, great promise for future research.

Since 1870, the central argument for common descent, the argument I've been concentrating on here, has been applied to an enormous number of natural phenomena. Darwin had worked hard to resolve particular difficulties about distribution, wondering how plant seeds might be carried to distant islands and how the worldwide distribution of some Alpine plants might be explained. In his garden at Down, he performed experiments to determine how long seeds could survive in saltwater and still retain the power to germinate. Using his estimates, he was able to show that the measured speeds of currents often allowed for transport from continent to island. He claimed that the scatter of Alpine plants at high elevations around the globe should be seen as the result of climatic changes. As the earth cooled, Alpine plants were able to invade the valleys and spread on the slopes of mountains previously uncolonized. With warming trends, they were no longer competitive in the lower elevations, and only the populations at higher levels were able to survive. As our understanding of the earth and its history has increased in the past decades, particularly with the understanding of continental drift and with a more detailed picture of past climates, Darwin's own answers have been refined and extended. The distribution of plants and animals around the world is explicable as the product of dispersal, descent, and modification. From the perspective of independent events of creation, it is simply a rigmarole.

The most powerful subsequent development of Darwin's central argument comes, however, from biologists' increasing ability to investigate the relationships among organisms at an ever finer grain. From the twentieth- or twenty-first-century perspective, descent involves the transmission of genes—segments of DNA in almost all organisms—and descent with modification occurs when the genes are altered or their organization is rearranged. A particular stretch of DNA is identified by the order in which four types of molecule, the *bases* or *nucleotides*, adenine, cytosine, guanine, and thymine (A, C, G, T, for short) occur along it. Modifications to a DNA segment can occur in different ways: one of the bases may be changed (a T replaced by a C, say); some of the bases may be cut out, or another segment of DNA may be inserted; sometimes parts of the segment may be inverted. (Famously, DNA molecules take the form of a double helix, in which the bases jut inside from the helical backbones, subject to the rules of pairing that As go with Ts, and Cs with Gs. DNA replicates by separation of the helices, with each strand serving as a template for the formation of a new partner. Many of the modifications occur when there are copying errors in this process.) With the advent of techniques for reading the genetic sequences in members of different species, the stage is set for comparison. Molecular evolutionists can construct an explanatory picture of the changes that have produced the genetic diversity we find.[35]

I shall have much more to say shortly about the ways in which our embryonic knowledge of genetic similarities and differences enriches the central argument of the *Origin* and vindicates the repudiation of Creationism in the 1860s. First, however, it's valuable to understand how this is the culmination of a process of comparative biology that starts with the most evident features of organisms—like the common bone structures in mammalian forelimbs that Darwin cited—and proceeds to ever more minute details. The striking affinities of anatomy and physiology point clearly to relationships, but the exact character of the relationships, the determination of just which organisms are closer kin, depends on the history of modifications of the genetic material.

Early in the history of classical genetics, long before biologists knew that genes are segments of DNA, they discovered that the hereditary material was carried on the chromosomes —long, threadlike structures in cells, that stain in distinctive ways (hence the name) and that can be seen under the microscope. Later in the twentieth century, biologists recognized that some chromosomes display characteristic banding patterns. Later still, they could make pictures of the chromosomes found in putatively related species—species that were seen as kin on the basis of the kinds of observable similarities Darwin used. They could then both demonstrate the affinities of chromosome number, chromosome size, and banding pattern, and use their analysis to uncover details of relationship that the cruder anatomical and physiological criteria could not determine.

The chromosomes are the bearers of the heredity material, and the banding patterns reveal the large-scale ways in which the genes are organized (each band contains many genes, and the finer grain of internal organization is not shown in the banding patterns). By investigating the chromosomes, biologists obtained a first, albeit rough, picture of the genetic similarities among organisms. One particularly interesting comparison, achieved in the second half of the twentieth century, looked at chromosomes from human beings and from chimpanzees, gorillas and orangutans.[36] Chimpanzees and the other great apes have 24 pairs of chromosomes, human beings 23 pairs. (See Figure 3.1.) With a single exception, the chromosome conventionally labeled "2" (the second longest), all our chromosomes can be matched, for size and for banding pattern, with chromosomes from the apes, and most closely with those of the chimpanzees. When the exceptional case, human chromosome 2, is set against the two unmatched ape chromosomes, it becomes apparent that the total length of those two chromosomes is almost exactly the same as the length of human chromosome 2—the single human chromosome is shorter by a very small amount—and that the banding patterns yield an almost exact match. The obvious Darwinian explanation for the similarity is that the four species descend from a common ancestor, with 24 pairs of chromosomes and that two of the chromosomes fused to form human chromosome 2.[37] If human beings were a completely separate creation, why did the creative force find it apt to form our species in the chromosomal image of the great apes?

Genes play an important role in the development and maintenance of organisms by providing the instructions for building proteins, the molecules that are omnipresent in the chemistry of cells. Similarities in genes, that is in DNA sequences, are reflected in similarities in proteins, and, in the past decades, as it has become ever easier to analyze proteins associated with particular biological processes and activities, biologists have been able to compare the structures of proteins in different species. Here again they discern the same story. The affinities Darwin recognized at the level of anatomy and physiology are supported by the relationships among the proteins. Organisms previously claimed to be more distantly related have proteins that differ more than those seen as closer kin. Moreover, as with the

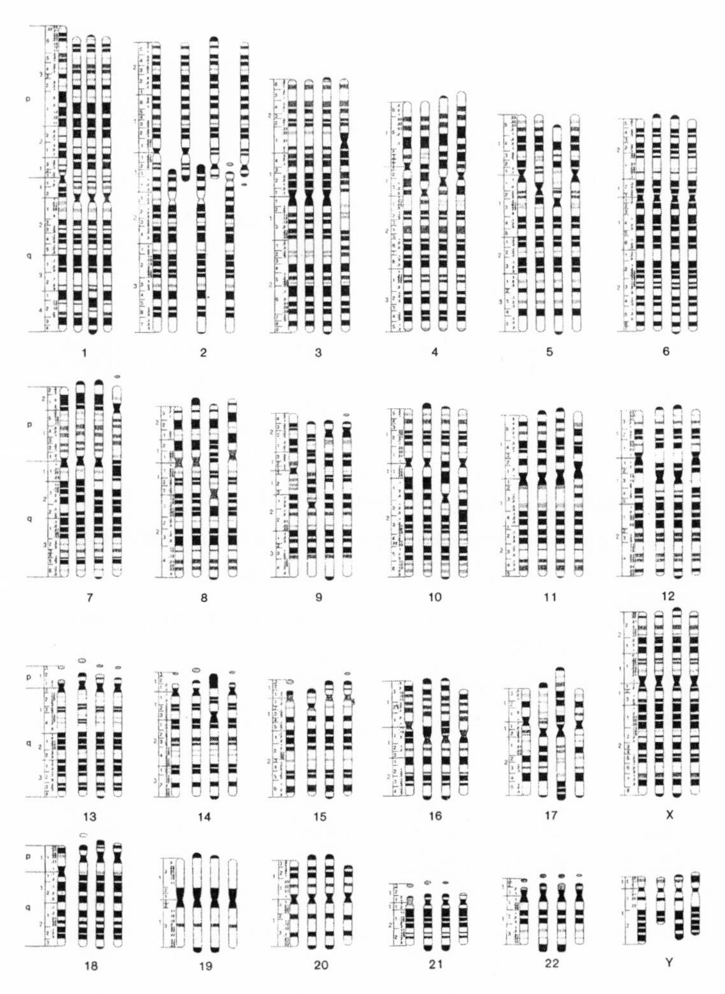

Figure 3.1 Comparison of the chromosomes of human beings, chimpanzees, gorillas, and orangutans. The chromosomes are shown in this order, from left to right, in each case.

chromosomal studies, the comparison of proteins provides the basis for a finer analysis, for the delineation of specific relationships that could not have been determined on the basis of gross anatomy. Occasionally, there are corrections, cases in which one species, supposed on the basis of anatomical comparison to be closer to a second rather than to a third, turns out to be more nearly related to the third. Even when this occurs, however, there is no basis for questioning broader conclusions about kinship. Nobody doubts that all three species are related more closely than they are to a vast number of more remote outsiders. Rather, the more fundamental criterion of modification at the genetic level (supposed to be reflected in the protein comparisons) is used to refine judgments arrived at by comparative anatomy and physiology.

Contemporary molecular genetics can now do what seemed, even a decade ago, improbably ambitious. Biologists, or, more exactly, their technicians and their robots, can produce reams of genetic sequence data that can be used to make still more refined comparisons among species. The recent sequencing of the genomes of human beings and of chimpanzees has demonstrated just how close[38] we are. There is an extraordinary overlap in the DNA sequences, reinforcing the picture obtained from the study of chromosomes. The principal differences seem to lie in those parts of the DNA that direct the rates at which particular genes are "switched on" to make their characteristic protein products. Yet this is only part of a much larger story, one that reveals how some genes present in organisms very unlike the animals we know and love—single-celled organisms like yeasts for example—have versions (segments of DNA with significantly similar sequences) in all sorts of organisms, fungi and ferns, worms and mollusks, fish and birds and mammals. The Darwinian explanation is that these genes originally evolved to direct some basic tasks within the cell—they are "housekeeping" genes—and multicellular organisms have taken over this machinery, sometimes modifying it, sometimes retaining parts of it that no longer serve their original purpose.[39]

Why should organisms so diverse share related DNA sequences if large groups of them have been separately created? The creationist response must be that these genes are an especially good design idea, that versions of them are needed to carry out important tasks in all these organisms—for if that is not so, then assigning them across the board is another piece of creative whimsy. That response falls afoul of the fact that some of the genetic machinery needs to be suppressed in many instances, to be regulated so that its presence doesn't interfere with important intracellular processes. Instead of being a brilliant plan on the part of "creative Intelligence," some of the originally useful genes turn out to be parts of the equipment that other organisms are stuck with, potential obstacles to be overcome.

If you were a talented engineer designing a whale from scratch, you probably wouldn't think of equipping it with a rudimentary pelvis. If you were designing a mammalian body, you might try to set things up so that development doesn't lead to a tangling of reproductive and urinary tubes so that one sex—females among marsupials like kangaroos, males among placental mammals (most mammals, including our own species)—is burdened with hernias waiting to happen. If you were designing a human body, you could surely improve on the knee. And if you were designing the genomes of organisms, you would certainly not fill them up with junk.[40]

The most striking feature of the genomic analyses we now have is how much apparently nonfunctional DNA there is. Birds have it, bees have it, we have it, but some simpler organisms, bacteria, for example, have far less of it. The genomes of multicellular organisms are full of what look like the residues of sequences that were once functional genes, often in many copies, but that have now become degraded and play no role in generating proteins. Even worse, there are long repeating sequences, some of which can increase beyond the point of simply being useless. Some of the worst disruptions of human functioning, both in early development (fragile X syndrome) and in forms of dementia (Huntington's disease), occur as the result of repetitive DNA that has been lengthened in the transmission from parent to child. From the Darwinian perspective all this is explicable—the molecular equivalent of the tinkering that is pervasive in the history of life at the anatomical level. Evolution works with a common bone structure to yield here a bat's wing, there a porpoise's paddle. It works with inherited DNA sequences to cobble together molecular solutions to the problems of arranging profitable biochemical processes. Over the history of life, the residues of past tinkering accumulate in the genome, the once-functional sequences, the degraded remains of genes, the long repeats. What rival explanation can creationists provide? They can't say these are strikingly good designs. For much of what we find is a disorderly botch, some of it dangerous and needing newly contrived methods of control. As the evidence accumulates, creationists increasingly must take refuge in responses Darwin saw as unsatisfactory evasions, appealing to the thought that these properties of life are unfathomable mysteries.

* * *

Despite the massive evidence for common descent, for a single tree of life, novelty creationism still lurks in the shadows of the intelligent design movement. The position is adjusted to avoid some Darwinian objections. Canny advocates will concede that some species are related by common descent, or even that this is the normal case in the history of life, but they insist that there have been special moments in that history when genuinely new designs have been produced by a creative agency. In effect, they draw bars across the allegedly unique tree of life, marking those places where the tree is broken, where history makes a fresh start.

At which places, exactly, is the continuity of life disrupted? Perhaps Darwin's opponents take more seriously his description of the origin of life—life "breathed" into the original forms—more seriously than his successors, or probably Darwin himself, would be inclined to do. Another prime example would be the origins of multicellular organisms. Yet another instance of the intervention of creative agency might be the appearance of our own species (an instance that has obvious religious resonances).

There are hints in *Of Pandas and People*, the textbook written to give a "balanced" presentation of issues about "origins," that the authors believe in an "abrupt appearance" of human beings.[41] If, as I suspect, their ambiguous language reflects a desire to break the tree of life, to separate *Homo sapiens* from everything else, then the intelligent design movement is committed to novelty creationism. Of course, if the intelligent design-ers were explicitly to assume this commitment, they ought to come to terms with the massive body of evidence linking human beings to the rest of nature—the common anatomical and physiological traits, the chromosomal similarities, the protein affinities, the DNA sequences we share with yeast, those we share with other vertebrates, those present in other mammals, the especially close kinship of our DNA with that of chimpanzees, the residues of once-functional sequences, the load of apparent junk. All this would have to be explained as the product of a special creative act, one that deserves to be labeled "intelligent." Contrary to appearances, it must be an especially bright design idea.

Contemporary novelty creationists don't take on that explanatory burden. Instead, they pursue a version of the strategy deployed by Genesis creationists: the best defense is a good attack. Just as the Genesis creationists attempt to argue that there are some stratigraphical anomalies, some places that show human and dinosaur remains in the same deposits, some signs of cataclysmic events, so too novelty creationists scour the scientific literature for unsolved puzzles. They raise familiar problems about the absence of transitional forms in the fossil record (one of the difficulties Darwin openly acknowledged and tried to address—the other is the efficacy of natural selection, a topic that will occupy us in the next chapter). And they allege that molecular analyses are at odds with Darwinian claims. As with Genesis creationism, a patient response would review all the accusations, expose misunderstandings and fallacious arguments, distinguish the unsolved problems from the genuinely unsolvable. Here too, however, I think a less conciliatory reply is also appropriate.

Let us concede, for the sake of present argument only, that the novelty creationists have fastened on real difficulties. Darwin's contemporaries knew very well that his proposals could not explain all the phenomena of the organic world.

Given the Lyellian alternatives, they opted overwhelmingly for the tree of life instead of events of special creation because the former, unlike the latter, explains something. How do matters stand a century and a half later? The answer is that, in breadth and depth, the capacity of the idea of descent with modification to explain the characteristics of living things has expanded enormously. By contrast, Creationism is just where it was in 1870 (or in 1859 for that matter), unable to advance a serious account of any of the kinds of phenomena on which Darwin built his case. Even if all the difficulties raised by novelty creationists were genuine, the choice between the Lyellian options should remain the same.

My judgment may appear harsh. Aren't there some triumphs for novelty creationism, some places where it faces up to the task of explaining the phenomena on which the Darwinian case for a single tree of life is based? Occasionally there are gestures that promise explanations of some of the evidence for the Darwinian conception of one tree of life. Consider the following passage from the recommended school text, *Of Pandas and People*:

> Darwinists take the widespread occurrence of fundamental chemical building blocks, processes, and organization as evidence of the common ancestry of life through macroevolution. Doesn't it make sense, say Darwinists, that the best adapted fundamental elements of ancient life forms would continue to be utilized by living systems, in spite of other changes through time?
>
> One response of design proponents is to cite the importance of the food chain. Food chains provide the support that makes possible large and diverse ecosystems. . . . Design proponents reason that common building blocks are a matter of efficiency. If

> the molecular building blocks used by predators were different from those of their prey, how could they utilize their food?[42]

New acts of creative activity generate organisms with the same fundamental mechanisms found in previous living things, so that the new ones can interact with the old ones—specifically so that they can eat, or be eaten, by them. At first sight, it's a plausible idea, about as plausible as the thought that the birds could evade the flood longer than the fish.

On further thought, there are obvious questions. Just which of the shared features cited in support of common descent are to be explained: the conserved DNA sequences, the junk that litters the genome, the similarities of chromosome number and banding patterns, or just the common metabolic pathways, cited by the text just before the passage I have quoted?[43] The text waves vaguely in the direction of a host of phenomena without offering any detailed account. Moreover, it mischaracterizes the Darwinian case for common descent. The point isn't that the continued use of "the best adapted fundamental elements of ancient life forms" would "make sense," but that more recent organisms are sometimes stuck with these even when they no longer help in a new environment.

This assessment, however, doesn't penetrate to the root of the trouble, which is that no genuine explanation has really been given. There are similarities in the basic mechanisms and processes used by all organisms—but also differences. Animals, after all, don't go in for photosynthesis. So the details matter. Just which features of past life would novel creation have to preserve if the newly created organisms were to interact with the old ones? The question is impossible to answer, for we have no idea about the powers of this creative agency. Would it be able, for example, to create organisms with molecular processes employing amino acids not found in the rest of the organic world? Initially, this might seem impossible, but the biochemical tricks of life are sufficiently various to generate doubt: perhaps Intelligence can introduce mechanisms that enable the new organisms to assemble the amino acids they need.

We don't know what Intelligence can or cannot do—and, equally, we have no clue about how it is directed in the process of creation.[44] It is easy to slide into illicit personification as you read *Of Pandas and People*, to suppose that Intelligence wants to produce "large and diverse ecosystems." Strictly speaking, of course, this is not the way we should be talking. Instead we should say that Intelligence is directed toward producing ecosystems of this sort. Apparently, Intelligence is also directed toward doing this by producing organisms that eat other organisms—and this, in itself, is a puzzling, even disturbing thought, since it would seem possible to equip all organisms with a device like photosynthesis that would avoid the messiness of predation. Intelligence also seems directed toward generating organisms whose needs can be met. Nonetheless, as phenomena of extinction make plain, it's compatible with the working of Intelligence that the needs of types of living things sometimes cease to be met—even in dramatic ways (as the mass extinctions, the most famous of which involves the demise of the dinosaurs, reveal).

Like other expositions of intelligent design, *Of Pandas and People* simply assumes that Intelligence is directed in various ways and that it has particular powers, without making explicit any precise principle about the direction or extent of the powers. We don't know how, and to what extent, it will be directed toward meeting the needs of organisms, or to producing diversity, or what kinds of interorganic relations it is to favor. Nor have we any conception of what powers and limitations Intelligence has. In consequence, there's no basis whatsoever for supposing that Intelligence would create new organisms with the features of those already around—or for determining which features would be preserved. No explanation is on offer, and there's no real engagement with the mass of Darwinian evidence for common descent.

* * *

Novelty creationism is far more comfortable when it hews to the traditional path of anti-Darwinism, that of offering criticisms rather than providing positive accounts of the phenomena. One favorite ploy is to claim that the molecular analysis of proteins confounds Darwinian expectations. So, for example, the school text *Of Pandas and People* offers a table comparing the differences in versions of a common protein, *cytochrome c*, across a range of organisms. (See Figure 3.2.) After noting that the variations are "not surprising," since they conform to "traditional taxonomic categories," the authors proceed to their confutation of Darwinism:

> The reason this finding is so surprising is that it contradicts the Darwinian expectation. As we move up the scale of evolution from the silkworm moth, that expectation (although it was probably never stated as a prediction) was to find progressively more divergence on the molecular level. This expectation holds true, even though we are comparing what may be described as contemporary representatives of progressively appearing

	Pig	Turtle	Frog	Tuna	Silkworm
Pig	0	9	11	16	25
Turtle		0	10	17	26
Frog			0	14	27
Tuna				0	30

Figure 3.2 Percentage differences among *cytochrome c* molecules found in various organisms (more exactly: percentage difference in amino acid sequences of *cytochrome c*). This is a simplified version of the table given in *Of Pandas and People.*

> classes, rather than descendants and ancestors. . . . Indeed, when comparing living organisms, Darwinism would predict a greater molecular distance from the insect to the amphibian than to the living fish, greater distance still to the reptile, and greater than that to the mammal. Yet this pattern is not found.[45]

As the table shows, the divergences between *cytochrome c* in silkworms, on the one hand, and tuna, bullfrogs, snapping turtles, and pigs, on the other, are all much the same. The differences in composition are 30, 27, 26, and 25 percent respectively.

There is a very good reason why the "prediction" was never stated explicitly, a reason that has no basis in the caution or duplicity of Darwinians—for it's not a "Darwinian expectation" at all. The principle that derives from the idea of a single tree of life is both simple and obvious. Molecules in organisms that have had a more recent ancestor in common will diverge less than molecules in organisms whose last common ancestor is more remote. It follows that the molecular divergence between tuna and turtle should be less than the molecular difference between tuna and silkworm—as indeed it is (17 versus 30 percent). Similar conclusions are upheld for other genuine consequences of the principle (that the divergence for pig

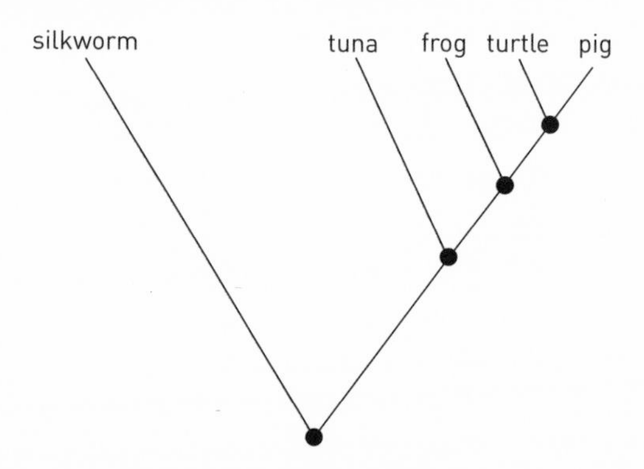

Figure 3.3 Tree diagram showing evolutionary relationships among five types of organisms. The heavy black circles represent the common ancestors of the organisms connected with them by upward-pointing lines. The lengths of those lines represent, very roughly, the time elapsed since the last common ancestor.

and bullfrog should be less than that for bullfrog and silkworm, for example). What does the principle tell us about the relation of the divergence between tuna and silkworm to the difference between pig and silkworm? Absolutely nothing. For, on the Darwinian conception of the unique tree of life, the most recent common ancestor of tuna and silkworm was also the most recent common ancestor of pigs and silkworms. That ancestor also lived a very long time ago. More recently pigs and tuna had their last common ancestor, and, since then, their versions of the protein will have diverged. (See Figure 3.3.) The alleged confutation evaporates.

Turn now to the complaints about the absence of "intermediates" in the fossil record. Paleontologists have reconstructed all sorts of sequences of fossils to show various transitions in the history of life, and canny novelty creationists acknowledge this. They are not committed to denying that descent with modification ever occurs, or even that it is sometimes caused by natural selection. Indeed, they should be quite unperturbed by the complaint, often launched by scientists against intelligent design, that their ideas are dangerous because they do not allow for the evolution of pathogens. Instead, they dispute the major transitions in evolution, the points at which genuine novelties emerge—birds or mammals, say, or the first land-dwelling animals, or even the first multicellular organisms. In none of these instances, they claim, can Darwin's defenders point to a fine-grained series of ancestral forms out of which the new types gradually emerge.

Darwin lamented the poverty of the fossil record, comparing it to a book in which most pages have been destroyed and, on the surviving pages, most of the words blotted out.[46] To be fossilized at all, an organism has to die in the right place at the right time under the right conditions, and even then there are good chances that subsequent events will destroy the traces. From the 1860s on, Darwinians have pointed to a reptile-bird intermediate, the famous *Archeopteryx*, to show that, despite the vicissitudes of fossilization, there are cases in which major transitions can be found in the rocks. Prominent novelty creationists demur, thus Phillip Johnson states, "If we are testing Darwinism rather than merely looking for a confirming example or two, then a single good candidate for ancestor status is not enough to save a theory that posits a worldwide history of continual evolutionary transformation."[47]

Johnson assumes that a fine-grained sequence of intermediates should be found in the fossil record. That, however, is something on which studies of the physics and chemistry of fossilization can, and do, pronounce. Estimates derived from these independent areas of science reveal that the chances of fossilization are so low that we should not expect to find intermediate forms in the profusion Johnson requires.

Yet paleontologists are sometimes lucky, as in the case of *Archeopteryx* and as in another example of a major transition, the reptile-mammal transition. Here there are many specimens of therapsids (mammal-like reptiles) and of early mammals. Johnson shows his appreciation of this case by referring to it as "the crown jewel of the fossil evidence for Darwinism."[48] He continues with an unusually accurate condensation of the paleontological results. "At the boundary, fossil reptiles and mammals are difficult to tell apart. The usual criterion is that a fossil is considered reptile if its jaw contains several bones, of which one, the articular, connects to the quadrate bone of the skull. If the lower jaw consists of a single dentary bone, connecting to the squamosal bone of the skull, the fossil is classified as a mammal."[49] It might seem very difficult for any animal to be "intermediate" between reptiles and mammals, given this criterion in terms of jaw morphology. There is, however, a very rich set of fossils showing reduction of the reptilian features and development of the mammalian traits. Particularly remarkable are fossils, most famously *Diarthrognathus* (which literally translates to "double jaw joint"), in which both types of jaw joint are present.

Johnson recognizes that this atypically rich collection of specimens has been preserved, but he denies that it provides what Darwinians need. "Creatures have existed with a skull bone structure intermediate between that of reptiles and mammals, and so the transition with respect to this feature is possible. On the other hand, there are many important features by which mammals differ from reptiles besides the jaw and ear bones, including the all-important reproductive systems."[50] It appears to be a reasonable demand. After all, the Darwinian argument relies on assuming that other mammalian features will correlate with dentary-squamosal jaw joints, and other reptilian features will go along with quadrate-articular jaw joints—as they do in living mammals and living reptiles. Yet that is just an assumption, and maybe things were different in the past. So we don't have really solid evidence for the transition unless we have fossils that show changes in "the all-important reproductive systems."

Once again, Johnson has described just the evidence that would satisfy him—and very cleverly, too. For, as he ought to know, there isn't a single mammalian reproductive system, but three—that of the *monotremes* (egg-laying mammals like the duck-billed platypus and the echidnas), that of the marsupials, and that of placental mammals. Moreover, as he also surely knows, the fossil record will virtually never reveal differences among any of these and the reptilian reproductive system, because the differences depend on the soft parts, parts that almost never get fossilized.[51] So the demand is for fossils of a very specific sort, and the physics and chemistry of fossilization tell us that the probability that those fossils will exist is effectively zero. It's precisely because our only access to the full range of characteristics that changed in the reptile-mammal transition is through the distinctive bone structures that can be preserved in the rocks, that paleontologists use the structure of the jaw joint as a diagnostic tool.

There are signs that Johnson recognizes that not all is well with the line of argument I've been examining, for he follows it up with an alternative. Darwinians are troubled by an *embarras de richesse.* "But large numbers of eligible candidates are a plus only to the extent that they can be placed in a single line of descent that could conceivably lead from a particular reptile species to a particular early mammal species. The presence of similarities in many different species that are outside of any possible ancestral line only draws attention to the fact that skeletal similarities do not necessarily imply ancestry. The notion that mammals-in-general evolved from reptiles-in-general through a broad clump of diverse therapsid lines is not Darwinism. Darwinian transformation requires a single line of ancestral descent."[52] This is a mass of confusions. Contemporary Darwinians think of the tree of life as bushy—at moments of major transition there are often many branches (branches on branches, and so on), most of which are short lived. Darwinians expect that there will be many closely related species—daughters and sisters and cousins and aunts. But it may be difficult, with the tools available, to reconstruct which of the clearly related species is ancestral to which, to sort out the mothers and the daughters from the sisters and the cousins and the aunts. The fossil record doesn't reveal the emergence of mammals-in-general from reptiles-in-general (whatever that would mean). Rather it shows that there were lots of related therapsid species that are potential ancestors to the first mammals, and we cannot tell (at least at present) which of them was the true progenitor.

Because novelty creationist attacks seem so often to be based on misunderstandings or unwarranted assumptions, the patient response requires genuine patience. Furthermore, it is easy, when working through the deficiencies of some criticism, to feel that the task is Sisyphean, that no sooner has this particular stone reached the summit than another will appear and require the same time-consuming work. Creationist literature is especially creative in its misunderstandings, in pinning odd views on the opposition and then playing "Gotcha!"—as when Darwinism is committed to a "prediction" about *cytochrome c*, or when it is required to exhibit fossils that show changes in the reproductive system. It isn't that Darwinism is a "flawed theory," beset by innumerable problems. Rather, creationists can manufacture spurious "problems" faster than evolutionists can unmask their sophistries.

Because of this, it's important to appreciate the less patient response, to understand the fundamental structure of the debate. As Lyell saw so clearly, once you admit that there have been different types of organisms on the earth at different historical stages, there are just two possibilities. Either the new ones come from the older ones, or the new ones spring into being from a new creative act. Darwin's long argument showed that Lyell had picked the wrong option. Recognizing a single tree of life can account for innumerable details of the organic world that Creationism can only regard as the whimsy of Intelligence—and Lyell himself appreciated the point. With the explosion of detail in the century and a half since, coupled with the continued explanatory bankruptcy of the creationist program, intellectual honesty requires that one follow Lyell's honorable lead.

Chapter Four

AT THE MERCY OF CHANCE?

> Some . . . deceived by the atheism they bore within them, imagined that the universe lacked guidance and order, as if it were at the mercy of chance.
>
> St. Basil the Great
> (quoted by Pope Benedict XVI in discussing intelligent design)

Henry Charles Fleeming Jenkin, once professor of engineering at the University of Glasgow, is best remembered for his work with Lord Kelvin in laying the transatlantic cable. He probably never knew of the great compliment Darwin paid him in 1869, in a letter to Hooker, in which Darwin reflected on the review that Fleeming Jenkin had published two years earlier. "Fleeming Jenkin," Darwin wrote, "has given me much trouble, but has been of more use to me than any other essay or review."[53] The trouble concerned natural selection, for Fleeming Jenkin's analysis suggested that the accumulation of small variations under selection could not produce the effects Darwin had claimed.

The difficulty was focused by using an example whose unconscious assumption of imperial majesty and racial superiority is breathtaking—Fleeming Jenkin probably expected his readers to glow and agree:

> Suppose a white man to have been wrecked on an island inhabited by negroes . . . Suppose him to possess the physical strength, energy, and ability of a dominant white race . . . grant him every advantage which we can conceive a white to possess over the native . . . yet from all these admissions, there does not follow the conclusion that, after a limited or unlimited number of generations, the inhabitants of the island will be white. Our shipwrecked hero would probably become king; he would kill a great many blacks in the struggle for existence; he would have a great many wives and children . . . In the first generation there will be some dozen of intelligent young mulattoes, much superior in average intelligence to the negroes. We might expect the throne for some generations to be occupied by a more or less yellow king; but can any one believe that the whole island will gradually acquire a white, or even a yellow population . . . ?[54]

The argument assumes a particular view of inheritance, that different characteristics in parents are likely to blend in their offspring, and the appeal to "color" is, unfortunately, well chosen to make the point. For not only did Darwin lack any rival account of the mechanism of heredity, but he would have been hard-pressed to deny that, with respect to some traits, traits like skin color, blending of parental characteristics is often observed.

Fleeming Jenkin raised other troubles as well, using Lord Kelvin's estimate of the age of the earth to argue that natural selection could not possibly have produced the observed diversity of life in the time allotted. Here, too, Darwin had no effective answer, for any attempt to "speed up" the operation of natural selection would fall foul of the fact that, in human history, no example of "species transmutation" had been observed. Small wonder, then, that in the wake of two powerful objections, one questioning the theoretical coherence of selection as Darwin understood it, and one using all the authority of physics to derive the inadequacy of selection, many who accepted the principle of a single tree of life sought different mechanisms for evolution. Some—the first champions of intelligent design, perhaps?—supposed that evolution had unfolded the diversity of living things, but that it had been guided throughout by the hand of the Creator.

* * *

It took several decades to appreciate that these apparently powerful arguments were wrong. Eventually, Kelvin's seemingly impeccable estimates were undermined by the discovery of radioactivity, and the age of the earth was fixed at over four billion years,[55] instead of a few hundred million. Eventually, the commonsensical notion of blending inheritance was replaced by a theory that identified discrete units of inheritance, as Mendel's insights were refined by T. H. Morgan and his successors. Even then, however, it took time to understand just how to marry Darwin's ideas with the new genetics, and to articulate a "genetical theory of natural selection."[56]

In the 1930s, however, the marriage was officially recognized, and natural selection was made respectable at last. Furthermore, the new genetic perspective revealed that many of the alternative causes of evolutionary change, proposed in the decades since 1870, could not do what was claimed for them. Almost by default, natural selection emerged as the principal mechanism of evolutionary change.

The dominant perspective, from the "modern synthesis" achieved in the 1930s to the present, takes evolution fundamentally to consist of changes within the genome—the replacement of some *alleles* (forms of a gene) with others, the insertions of extra copies of a gene, or the rearrangement of the genetic material. Localized variation can occur in three distinct ways. There can be replacement of one of the DNA bases (A, C, G, T) by one of the others (*substitutions*—as, for example when a stretch of DNA consisting of the sequence CGGACTCG is replaced by CGGA**A**TCG). There can be deletions of part of the sequence (CGGACTCG gives way to CGGTCG). Or there can be insertions (CGGACTCG becomes CGG**G**ACTCG). Most of these changes disrupt the normal functioning of the genetic regions in which they occur. Frequently, the proteins that would usually be formed are no longer generated, or are generated in variants that fail to support the processes for which they are required. Occasionally, however, a mutation will be advantageous, enabling the organism in which it appears to do better in the competition for survival and reproduction. When that occurs, natural selection works in very much the way Darwin conceived of it.

More global changes may well prove to be critical to the fashioning of advantageous genetic novelties. Perhaps the principal steps in the evolution of life involve the miscopying of particular regions of DNA, fruitful errors that allow for duplications of genes or for reshuffling of the genetic material. If a modification like this juxtaposes two DNA segments that were previously relatively far apart, the result may be to alter the frequency with which proteins are made—perhaps a region that directs the formation of a protein is now equipped with an adjacent sequence that increases (or decreases) the rate at which that protein is produced. Comparative genomic sequencing, which has already begun and that is likely to provide enormous information in the next decade or two, can enlighten us about the ways in which the most important evolutionary changes, the traits that lead to significantly different forms of life, have originated. Even if the source of a great leap forward lies in the rearrangement of genetic material rather than a new local mutation, the initial advantage it bestows will fuel its spread by natural selection, just as Darwin envisaged, and may thereby serve as the basis for the next advance.

Central to Darwinian orthodoxy is the idea that these fundamental changes in the genetic material are not directly caused by the needs of the organism. The fact that a mammal would survive longer or reproduce more if it had a thicker coat does not mean that the genes causally involved in the production of hair and fur will mutate more frequently, or that, when they do, the variations will tend toward increased coat thickness. Similarly, our understanding of the ways in which the *gametes* (sperm and eggs) are formed in many organisms[57] forecloses the possibility that characteristics acquired during an organism's lifetime can be passed on to its descendants, scotching the proposal that animals' efforts to meet their needs can accelerate the pace of evolutionary change. Hence two of the mechanisms heralded by earlier evolutionists—including Darwin himself—as alternatives to natural selection were discredited by the modern synthesis of the 1930s.[58]

Contemporary evolutionary theory doesn't provide only a theoretical account of how evolution under natural selection would occur, but also investigates the ways in which natural selection produces changes in particular populations. This is done most readily for microorganisms—bacteria forced to grow in harsh environments, on media that fail to supply some crucial requirement for their survival. In the past 70 years, laboratory studies of this sort have demonstrated natural selection in organisms with relatively short generation times, and biologists have been able to measure rates of mutation. There have also been studies of natural selection in the wild, perhaps most spectacularly the detailed investigation of the birds that, according to legend, inspired Darwin to think about evolution in the first place. For over a quarter century, Peter and Rosemary Grant have led a research team that has thoroughly studied the finches on several islands in the Galapagos archipelago, showing through detailed observation of the birds in successive generations, how natural selection has modified them, most notably in the size and shape of the beak.[59]

Another famous study, one that will be relevant later, concerns the persistence of sickle-cell anemia. This disease arises in people who have two copies of a mutant *allele* (a variant form of a gene), the *S* allele, which generates an abnormal form of hemoglobin. Because the abnormal molecules have a tendency to form rigid crescent shapes under conditions where oxygen is in relatively short supply, these molecules can become stuck in narrow capillaries, leading to problems in blood circulation. Patients then experience crises, they suffer from kidney failure and strokes, and they often die young. You might think that natural selection would tell against the *S* allele, and that the normal version, the *A* allele, would be virtually universal. That hasn't happened because the mixed combination, *AS*, not only yields the benefits associated with normal hemoglobin, but also provides protection against malaria. So, in African populations, and in other areas in which malaria is widespread, the *S* allele persists at a low but stable frequency, one that the theory of natural selection can predict quite precisely.

Tracking down the exact way in which natural selection is operating on a wild population of organisms—that is, a population outside the controlled conditions of the laboratory—is extremely time consuming, for researchers have to consider alternative potential explanations for the increase of the trait that seems to be favored. They also have to recognize that genetic differences are often reflected in a number of distinct characteristics, for genetic modifications typically have multiple effects, and it would be wrong to fasten on one of these as the crucial change that confers an advantage, without looking at the entire package. Seven decades of patient investigation have produced an impressive array of examples, some more complete and rigorous than others, enough certainly to show that natural selection can do the sort of thing Darwin claimed for it. The only rival mechanism for evolutionary change that has been theoretically vindicated and empirically demonstrated is "genetic drift," a process in which equivalent alleles increase or decline in frequency in populations as a result of the accidents of mating or survival. Although this process may have helped to maintain genetic variation, it seems quite unlikely that the principal transitions in the history of life, many of which seem to show new, if clumsy, adaptations to a different environment, have involved lucky victories by some genes over equivalent alternatives. So Darwin's thesis that natural selection has been the main agent of evolutionary change has won vindication by attrition. It seems to be the only potential cause left standing.

* * *

Yet, despite everything that investigators of natural selection have accomplished—and their achievements should not be underestimated—there are plainly things they have no hope of demonstrating. Given the difficulties of field research, any well-supported conclusion has to be relatively limited in scope. You show that a particular trait in a species has been shaped or maintained by natural selection, that finch beaks adapt to the distribution of available food items, for example. Darwin's opponents can concede the studies, and question the significance of the results. This only shows, they declare, that natural selection can produce small-scale changes. It can replace one form of a finch with a different type. It's a far cry from that to think that natural selection can generate (say) land-dwelling animals from fish. Nor does it help to point to the laboratory, and explain how all sorts of metabolically ingenious bacteria or pesticide-resistant insects can arise through natural selection. The opponents will charge that this is just more of the same, evolution "within a kind," perhaps a bit more dramatic than the results from the field, but also potentially compromised by the fact that the process has occurred in a tightly controlled setting. They want something much grander, a detailed study showing natural selection "transmuting" one "kind" into another—giving an amphibian from a fish, or a bird from a reptile, for example.

Nobody can answer that demand. From a Darwinian perspective, however, that isn't because the theory of natural selection is seriously flawed, but because the demand is absurdly naïve. Biologists have measured mutation rates. They know that favorable variations arise by mutation quite rarely, and that, if they were to try experimenting on the natural selection of organisms with relatively long generation times, it would take the lives of thousands of successive investigators to provide even the slightest chance of even the first steps toward experimental success.[60] They know that the earth is ancient, that geological time has offered far more opportunities for evolutionary "experiments" than successive generations of human beings could manage. They know that there is just one tree of life, and that the evolutionary changes in question did occur in the expanses of geological time. On a human timescale, natural selection can be observed to do small things. Nothing else is known as a potential agent for the evolutionary changes. Therefore, it's reasonable to think that the action of selection will produce more dramatic changes over longer periods, that the very large changes would be observable on a timescale inaccessible to human beings, one vastly greater than that of even our most protracted squabbles.

Of course, if some of the pieces of "knowledge" I've attributed aren't well established, then there will be grounds for questioning this defense. This is one reason I've tried to show why contemporary science is committed to an ancient earth and a single tree of life. As we shall see, intelligent design allows for, perhaps even encourages, backsliding on these points, but the official doctrine is that there is no quarrel with them. According to that doctrine, even if Darwinians are right about the age of the earth and the relatedness of all organisms, there are reasons to doubt whether the successes of selection in the small automatically translate into successes with more significant evolutionary transitions.

At the heart of the intelligent design movement are two kinds of arguments, both designed to question the thought that natural selection "scales up," by identifying transitions that could not be managed by selection. The "concrete case" argument, as I shall call it, selects a collection of evolutionary changes, discusses them in detail, then endeavors to show that there's no conceivable process of natural selection that could have started from the original group of organisms and culminated in the finally modified group. The "computational" argument abstracts from the details of individual cases, and presents them in terms of a more skeletal description, which is deployed to assign some basic probabilities. This basic assignment then allows for a calculation of the probability that the transition could have come about through the action of natural selection, and, since the estimate is extraordinarily tiny, the conclusion is that causation by natural selection is, to all intents and purposes, impossible.

Neither of these lines of argument is original with contemporary advocates of intelligent design. The concrete case argument is presented by Darwin himself as one of the "difficulties on theory" he attempts to address. The computational argument is prefigured in the writings of twentieth-century evolutionists who follow Darwin's lead in raising difficulties they intend to explore—and possibly resolve—by finding other natural mechanisms that would substitute for or complement natural selection. Precisely because anti-selectionism can be pursued, thoughtfully and respectably, within evolutionary theory, these prominent strands in the intelligent design movement are hard to dismiss as "not science." By the same token, however, it's reasonable to wonder why intelligent design casts itself as the source of a new alternative, one that demands a novel high-school curriculum, rather than simply as proposing that biology classes might be made more exciting by considering some of the suggestions of maverick evolutionists.

The answer is that intelligent design is a two-part doctrine. Despite the fact that its negative part, its anti-selectionism, occupies almost all the movement's writings, there's also a positive claim, the thesis that whatever cause produced particular changes in the history of life deserves the label "intelligent." Hence, two issues need to be addressed. First, how troublesome are the complaints, the versions of the concrete case and computational arguments? Second, even if we were to wonder whether natural selection can yield the outcomes to which the complainers point, what reasons are there for supposing that the actual cause, whatever it is, is intelligent?

* * *

Darwin's own consideration of the concrete case argument focused on some complex organs and structures that he rightly believed to be hard to understand in terms of natural selection. Two examples are prominent in the *Origin*, the eye and the electric organs found in some fish. The latter example disconcerted Darwin because the fish with electric organs are of very different types, and have their organs in different parts of their bodies. Much to his relief, research on electric fish carried out by a contemporary who was not sympathetic to evolutionary ideas—"McDonnell of Dublin (a first-rate man)"[61]—revealed that, for each type of fish with an electric organ, there is a related fish with a similar organ (not functionally electric) in the same position. What had initially appeared to be the challenge of understanding how different electric organs had been built from scratch, became the much simpler question of how a similar change had occurred in each instance.

Darwin himself offered a tentative proposal about the evolution of the eye. He supposed that sensitivity to light might come in degrees, and that it might be possible to find, among existing organisms, some with a crude ability to respond to light, others with a more refined capacity, and so on in something like a series. Perhaps, he speculated, research on these creatures might expose reasons why the different levels of sensitivity provided an advantage over rival organisms who had less, thus providing a way of answering (or sidestepping) the creationist quip, "What use is half an eye?"

It has taken more than a century of research on a wide variety of organisms to demonstrate that Darwin's hunch was basically right. Appearances to the contrary, organs and structures sensitive to light can be assembled piecemeal, with the intermediates enjoying some advantage over the competition. Biologists have studied organisms that respond to the light that impinges on their surfaces, organisms with indentations of the superficial layer that are able to acquire information about the direction of the light, organisms with deeper indentations whose light detection is more fine grained, organisms that have a structure resembling a pinhole camera, organisms that interpose a translucent medium between the surface and the aperture through which the light comes—and so on. By studying this sequence of organisms, they have been able to explore the transitions through which relatively crude abilities to detect light were successively refined.[62] One feature of the story deserves emphasis. Darwin didn't start with a comparison between the fully formed eye—in a human being or an octopus, say—and then think of the component parts as being introduced, one at a time. He resisted the challenge to explain first the advantage of an eighth of an eye, then the advantage of a quarter of an eye, and so on, and focused instead on a function, light sensitivity, that might have been refined from an initial state of absence. To put it more bluntly, he didn't allow his envisaged challengers to define the sequence of "intermediates" for him.

Savvy champions of the concrete case argument know this story. They appreciate Darwin's ingenuity in responding to the challenge, and, although they think the response ultimately fails, their reasons for this judgment depend on a more general problem for evolution under natural selection. That more general problem derives from the fine structure of the components of complex organs (like eyes), the molecular mechanisms that have to be in place for eyes to work. For all Darwin's cleverness, he failed to appreciate the full depth, and the full generality, of the difficulty confronting him.

The principal exponent of the complex case argument is Michael Behe, a professor of biochemistry, who argues at length in *Darwin's Black Box* that the real troubles of natural selection become visible when you appreciate the molecular components of complex biological systems. Almost everywhere you look in nature, there are complicated structures and processes, with many molecular constituents, and all the constituents need to be present and to fit together precisely for things to work as they should. Biochemical pathways require numerous enzymes to interact with one another, in appropriate relative concentrations, so that some important process can occur. If you imagine a mutation in one of the genes that directs the formation of some essential protein, or if you suppose that the genetic material becomes shuffled in a way that allows for differences in the rates at which proteins are formed, it looks as though disaster will ensue. Crucial pieces will be missing, or won't be present in the right proportions, so that everything will break down. How then could organisms with the pertinent structures or processes have evolved from organisms that lacked them?

Behe offers numerous instances of molecular machines that, he claims, could not have been built up in stages by natural selection. Among his most influential examples is a discussion of devices that some bacteria use for motion, flagella. He contrasts the bacterial flagellum with a different motor, used by other cells, the cilium. "In 1973 it was discovered that some bacteria swim by rotating their flagella. So the bacterial flagellum acts as a rotary propeller—in contrast to the cilium, which acts more like an oar."[63] Both flagella and cilia are intricate structures, and Behe describes the many molecular parts and systems that have to be present if they are to do their jobs. He concludes that the complexity of the organization dooms any attempt to explain its emergence as the result of natural selection. "As biochemists have begun to examine apparently simple structures like cilia and flagella, they have discovered staggering complexity, with dozens or even hundreds of precisely tailored parts. It is very likely that many of the parts we have not considered here are required for any cilium to function in a cell. As the number of required parts increases, the difficulty of gradually putting the system together skyrockets, and the likelihood of indirect scenarios plummets. Darwin looks more and more forlorn."[64] Indeed, the most famous portraits of Darwin hardly make him look exactly cheerful, but it's worth asking why examples like these should render him more forlorn.

Perhaps it seems obvious. Natural selection depends upon mutations that are not produced in response to the organism's needs. The bacteria are at the mercy of chance, which will fling in this variant protein or that, with negligible probability that the latest novelty will fit with what went before or will contribute to the design project of building a flagellum. In essentials, however, this is precisely parallel to an old creationist strategy, just the one that Darwin sidestepped in the case of the eye. Behe has specified how the intermediates are to be formed, and it isn't surprising that his preferred scenario has the air of impossibility.

What exactly is known about the bacterial flagellum? During the past few decades, careful molecular studies have identified the genes that direct the assembly of the motor, and have explored the ways in which it is put together in the development of an individual bacterium. Some mutations in these genes allow for bacteria to move, albeit less efficiently. What is currently missing, however, is that systematic study of the differences among bacteria with flagella and bacteria without that would parallel the knowledge attained in the case of vision. A sufficiently intensive study of the genomes of bacteria that lack flagella would enable biologists to explore the potential role of some of the crucial genes, and of the proteins they give rise to, when others are absent, and thus enable them to make more progress with Behe's apparently formidable challenge.

Most sciences face unsolved problems—indeed the exciting unsolved problems are the motivators for talented people to enter a field. Chemists still struggle to understand how newly made proteins fold into their three-dimensional shapes as they are synthesized. To take an instance closer to hand, Behe's own discussion acknowledges that there's still a lot to learn about the molecular structure and functions of cilia. Unsolved questions are not typically written off as unsolvable—nobody proposes that there's some special force, unknown to current chemistry (an "intelligent force" perhaps?) that guides the proteins to their proper forms, or some hand that assembles the cilium in the development of an individual bacterium. Why, then, should we believe that the problem of the bacterial flagellum is unsolvable? Just because, in the absence of systematic molecular studies of bacteria with and without flagella, we can't currently give a satisfactory scenario for the evolution of the bacterial flagellum under natural selection, why should we conclude that further research couldn't disclose how that evolution occurred?

We are beguiled by the simple story line Behe rehearses. He invites us to consider the situation by supposing that the flagellum requires the introduction of some number—20, say—of proteins that the ancestral bacterium doesn't originally have. So Darwinians have to produce a sequence of 21 organisms, the first having none of the proteins, and each subsequent organism having one more than its predecessor. Darwin is forlorn because however he tries to imagine the possible pathway along which genetic changes successively appeared, he appreciates the plight of numbers 2–20, each of which is clogged with proteins that can't serve any function, proteins that interfere with important cellular processes. These organisms will be targets of selection, and will wither in the struggle for existence. Only number 1, and number 21, in which all the protein constituents come together to form the flagellum, have what it takes. Because of the dreadful plight of the intermediates, natural selection couldn't have brought the bacterium from there to here.

The story is fantasy, and Darwinians should disavow any commitment to it. First, there is no good reason for supposing that the ancestral bacterium lacked all, or even any, of the proteins needed to build the flagellum. It's a common theme of evolutionary biology that constituents of a cell, a tissue, or an organism are put to new uses because of a modification of the genome. Perhaps the immediate precursor of the bacterium with the flagellum is an organism in which all the protein constituents are already present, but are employed in different ways. Then, at the very last step there's a change in the genome that removes whatever chemical barrier previously prevented the building of the flagellum. In this organism (the precursor), the function of one of the proteins is to increase the efficiency of a particular energy-transfer process. The precursor of the precursor lacked that protein, so that the genetic change that led to the precursor improved a process that was previously adequate. So it goes, back down a sequence of ancestors, all quite capable of functioning in their environments but all at a selective disadvantage to the bacteria that succeeded them.

Isn't this all fantasy too? Of course[65]—but it is no more the product of speculative imagination than Behe's seemingly plausible assumption that the components of the flagellum would have had to be added one by one, and would have sat around idly (at best) until the culminating moment when all were present. Moreover, we were supposed to be offered a proof of impossibility, and that won't be complete until Behe and his allies have shown that all the conceivable scenarios through which bacteria might acquire flagella are flawed. Really demonstrating impossibility—or even improbability—here and in kindred instances, is extremely difficult, precisely because it would require a much more systematic survey of the molecular differences among bacteria.

The serious way forward is to amend our ignorance, by sequencing the genomes of different bacteria, with and without flagella. Using our current knowledge of the genetic basis of the flagellum, researchers would be able to specify more clearly what the intermediate forms—those with some, but not all, of the crucial genes —might have been like, and what functions the relevant proteins might have served. Until we know these things, efforts to describe intermediates will be so much whistling in the dark. Behe's examples rely on guesses that simply anticipate what this hard work would reveal.

So we have the illusion of an impossibility proof. Allegedly there could be no sequence of intermediates concluding with the fortunate, flagellum-bearing bacterium, in which each member of the sequence enjoyed a selective advantage over its predecessor. Behe's story (quite charmingly told in *Darwin's Black Box*) offers his own preferred version of what the sequence would have to be like. Since Darwinians have no commitment to simpleminded stories of sequential addition of components, there is no reason to accept Behe's description. Because the same rhetorical strategy pervades his entire book, showing up in all the instances of the concrete case argument he provides, all the parade of examples really shows is that there are some interesting problems for molecularly minded evolutionists to work on, problems they might hope to solve in the light of increased understanding from comparative studies of the genetics and development of a wide variety of organisms.

* * *

The computational argument occurs in a variety of forms in current intelligent design literature, sometimes with relatively simple calculations of infinitesimal probabilities, on other occasions with much more technical specification of conditions under which we should make the "design inference" and conclude that some aspect of life has been intelligently designed.[66] Whether or not intelligent design-ers attempt to be fully explicit about the requirements for invoking design, all their versions require the preliminary step of arguing that it is highly improbable that various aspects of life on earth could have emerged through natural selection. To use an analogy much beloved by earlier creationists, Darwinian claims about selection and the organization of life are equivalent to the idea that a hurricane in a junkyard could assemble a functioning airplane.

Besides providing the concrete case argument, Behe offers several versions of its computational cousin. Here's his attack on a scenario for the evolution of a blood-clotting mechanism, tentatively proposed by the eminent biochemist Russell Doolittle:

> Let's do our own quick calculation. Consider that animals with blood-clotting cascades have roughly 10,000 genes, each of which is divided into an average of three pieces. This gives a total of about 30,000 gene pieces. TPA [tissue plasminogen activator] has four different types of domains. By "variously shuffling," the odds of getting those four domains together is 30,000 to the fourth power [presumably Behe means that the chance is one-thirty-thousandth to the fourth power], which is approximately one-tenth to the eighteenth power. Now, if the Irish Sweepstakes had odds of winning of one-tenth to the eighteenth power, and if a million people played the lottery each year, it would take about a thousand billion years before *anyone* (not just a particular person) won the lottery. . . . Doolittle apparently needs to shuffle and deal himself a number of perfect bridge hands to win the game.[67]

At first sight, this looks very powerful, since, given the time available for the evolution of life on earth (four billion or so years), it seems extremely improbable that the clotting mechanism could have evolved through natural selection.

Yet we should think carefully about the ways in which the pertinent probabilities are calculated. Behe is relying on two general ideas about probability. One is the thought that, when events are independent of one another, the probability that both will occur is the product of the individual probabilities—if you toss a fair die twice, then the chance of getting two sixes is 1 in 36; for, on each toss, the probability is 1 in 6, and, since the tosses are independent of one another, you multiply. The other idea is that, when you have a range of alternatives and don't have any reasons for thinking that one is more likely to occur than another, each of the possibilities has an equal chance. This idea, the notorious "principle of indifference," is known to be problematic, but, judiciously employed, it serves us well in some everyday contexts—as, for example, when we conclude that each side of the die has the same probability of falling uppermost.

Even in ordinary life, however, there are occasions on which applications of these principles would lead us to obviously unacceptable conclusions, so that we would rethink our computations. Consider a humdrum phenomenon suggested by Behe's analogy with bridge. You take a standard deck of cards and deal 13 to yourself. What is the probability that you get exactly those cards in exactly that order? The answer is 1 in 4×10^{21}. Suppose you repeat the process 10 times. You'll now have received 10 standard bridge hands, 10 sets of 13 cards, each one delivered in a particular order. Scrupulously, you record just the order in which all these cards were received, and calculate the chance that this event occurs. The probability, you claim, is 1 in $4^{10} \times 10^{210}$, which is approximately 1 in 10^{216}—notice that this denominator is enormously larger than Behe's 10^{18}. It must be really improbable that you (or anyone else) would ever receive just those cards in just that order in the entire history of the universe. But you did, and you have witnesses to testify that your records are correct. Excitedly, you contact Michael Behe to announce this quite miraculous event, surely evidence of some kind of Intelligence at work in the universe.

Your report would not be well received. Like everyone else, Behe knows how to understand this commonplace occurrence. Given the way in which the cards were initially arranged, the first deal was bound to go as it did. Given the shuffling that produced the ordering prior to the second deal, that deal, too, was sure to give rise to just those cards in that order; and so on. So there was a perspective, unknown to you, from which the probability of that sequence of cards wasn't some infinitesimally small number, but one (as high as chances go). If you describe events that actually occur from a perspective that lacks crucial items of knowledge, you can make them look improbable. We know enough about card dealing and coin tossing to understand this, and to see the calculation I attributed to you as perverse—for, although you don't know what the initial setup was, you should have recognized that there was some initial setup that would determine the sequence. Hence you should have known that application of the two general principles of probability in this context would provide a misleading view of the chance that this particular sequence would result.

In the case of the evolution of blood clotting, our ignorance is deeper. Not only do we not know what the initial molecular conditions—the prior state of the organisms in the population in which blood clotting emerged—were, we also don't know whether that initial state favored certain sorts of molecular changes rather than others. We have reason to think that Behe's assumption that there's a precise chance of 1 in 30,000 that each gene piece will participate in the "shuffling" process is incorrect. For, given what we know about mechanisms within the genome, the idea of exactly equal chances is suspect. But we don't know whether, given the initial molecular state, the chance of the cascade Doolittle hypothesizes remains infinitesimally small or whether it is actually one. Any estimate of the probability here is an irresponsible guess.

My imagined experiment with the deck of cards suggests a different way to think about the problem. Imagine that all the hands you were dealt were mundane—fairly evenly distributed among the four suits, with a scattering of high cards in each. If you calculated the probability of receiving ten mundane hands in succession, it would naturally be a lot higher than the chance of being dealt those very particular mundane hands, with the cards arriving in precisely that sequence (although it wouldn't be as high as you might expect). Blood clotting might also work in the same way, depending on how many candidates there are among the 30,000 "gene pieces" to which Behe alludes, that would yield a protein product able to play the necessary role. Suppose that there are a thousand acceptable candidates for each of the four positions in the molecule we need (TPA). The chance of success on any particular draw is now about 1 in 2.5 million. If there were 10,000 tries a year, it would take, on average, two or three centuries to find an appropriate combination, a flicker of an instant in evolutionary time.

My assumptions are no better—and no worse—than Behe's, for neither of us knows how tolerant the blood-clotting system is of the molecular combinations that the animals in question (whatever they were) might have supplied. We simply don't know what the right way to look at this problem is. But, given our ignorance, we shouldn't make wild guesses and then declare that the probabilities are so low that evolution under natural selection is impossible. A better research strategy would be to try to assemble information that will replace the guesses with serious estimates.

Moreover, even when the chance of a particular event turns out to be extremely small, it is important to resist the idea that that event could not have occurred. Imagine that you own a ticket in a lottery with an extremely large number of tickets—a million, say—and that the lottery is decided by a fundamentally random process, one that has no underlying causal basis by which the outcome will be determined. (You might suppose that each ticket is associated with a specific atomic nucleus of some radioactive element, and that the prize will go to the person whose nucleus decays first.) On any perspective we might justifiably adopt, there is a probability of one in a million that you will win, and similarly for all the other ticket holders (nobody has more than one ticket). Clearly, somebody will be lucky, and, after the fact, we'll have to admit that something very improbable has occurred. Moreover, this conclusion remains valid even if the number of tickets is vastly expanded. However small a probability we compute, it would be wrong to suppose that an outcome with that probability would be impossible. Finally, even if some people had massive numbers of tickets, it would still be possible for someone holding only one to win. That shows that an extremely improbable outcome, one much less probable than alternatives, is still possible.[68]

So there is another way to think about the allegedly minute probabilities of life as we know it. Our galaxy has roughly 200 billion stars, and the total number of galaxies, each with similar numbers of stars is—literally—astronomical. Some fraction of these stars has planets that are potentially suitable for sustaining life. Each such planet could be conceived as "buying a ticket" in an enormous lottery, with the prize, or one of the prizes, consisting in the emergence of life. For any of those planets, the probability of gaining that prize is extremely small, but if there are enough of them, one will win. That lucky planet may have been ours.

Because the computational argument is so pervasive in the intelligent design literature, it's worth looking at another version of it, one that focuses on a major challenge for Darwinism, the problem of the origin of life. In past decades, there have been many attempts to develop scenarios, or partial scenarios, for the emergence of the first living things. Along the way, there have been moments of excitement: the discovery that the building blocks of proteins (amino acids) could be generated from a soup of much simpler molecules (hydrogen, methane, ammonia, and so forth), molecules that might have been present in earth's primeval state; the discovery that there are forms of RNA that can form and replicate in a medium containing their basic constituents, and that can cut and splice other RNAs. Yet there are still extremely difficult problems in understanding how the molecular components of even the simplest cell might evolve, and how they might be assembled into a single package. It is thus no surprise that champions of intelligent design declare that the probabilities that simple living things could arise "by chance"—or, more exactly, by the mechanisms adduced by Darwinians—are so low that some quite different explanation is required.

This fertile territory breeds many versions of the computational argument. I shall focus on a representative calculation, that offered by Stuart Pullen in his *Intelligent Design or Evolution?* who tries to show that the assembly of the insulin protein within a primeval soup would be wildly improbable.[69] Pullen begins by using data from the experimental research of Stanley Miller, one of the principal figures in attempts to simulate the production of biologically significant molecules from soups of prebiotic constituents. Proteins are built up by selecting from a menu of 20 amino acids, and, although three of these were not synthesized in Miller's experiments, Pullen generously allows for them to occur at low rates.[70] He then uses a favorite analogy, imagining a "trapped scientist" who tries to find the combination of a lock by selecting blocks that represent the amino acids from a container (in this instance a truck!) that represent the biologically significant molecules formed in the primeval soup. If the scientist picks appropriate proteins in the right order, the lock opens, and he is free. The analogy assumes that the amino acids are found in the relative frequencies Miller reports (subject to the generous assumption that some molecules that aren't found occur at low rates), and that the formation of the protein, like the draws made by the trapped scientist, consists in a sequence of random, independent, trials. Pullen triumphantly reports his conclusion. "Odds of evolving in the primordial soup are 1 time in 2^{350} tries or 1 time in 2.2×10^{105} tries. This can never happen."[71]

What, never? Well, hardly ever. Yet Pullen's apparently generous estimate, like Behe's, depends on filling gaps in our knowledge with convenient assumptions. Why should one believe that the ratios obtained from Miller's experimental reports are good guides to the pertinent probabilities? What Miller hoped to demonstrate—and brilliantly succeeded in showing—was that amino acids could be synthesized from a mixture of prebiotic constituents. He did not claim that the mixtures used in his experiments were the only possible soups out of which amino acids might be generated, or that they were accurate facsimiles of the primitive conditions on the earth. Rightly so, for although we can propose hypotheses about how it might have been, nobody knows how it actually was. Miller's work is significant because it answers a particular question. Is it possible to obtain amino acids from a mixture of simpler molecules (like hydrogen, methane, and ammonia)? Pullen presses the data obtained into a computation about a different question. Is it possible to obtain insulin by a random process in a Miller soup? Now we don't know that Miller's specific soup is anything like the primeval mix. We don't know whether insulin was formed in it (nor is it at all clear how we might resolve that question). And we don't know if it would have been formed by some random process of molecules bumping into one another that might be like the sequential tries of the trapped scientist. Consequently, we have no idea whether the probabilities assigned to the insertion of particular amino acids are correct (they might be off by a factor of tens, hundreds, or thousands), no idea whether it is right to multiply them, and even no idea whether the task assigned the Darwinian is a significant one (if insulin wasn't made in the primeval soup, then the low probability is irrelevant).

The origin of life is a very hard problem precisely because we have so little idea about the constraints on a solution.

Darwinians assume that, from some unknown initial condition, biologically significant molecules—nucleic acids and proteins—would be generated and would eventually be assembled into a cell. The challenge is to go further, to specify how it might have been done. To respond to that challenge, you have to guess, to make assumptions about the initial conditions—and some inspired guesswork, followed by ingenious experimental research, has revealed that some aspects of original life can be simulated. What the decades of research also reveal is that our ignorance of those initial conditions is so extensive, and the range of possible assumptions so vast, that probability estimates are likely to be deceptive. Given a particular set of assumptions, guesses, you can calculate the probability that the evolution of life will proceed according to a particular scenario, but, because there are so many rival alternative guesses, you have no basis for assigning a probability to the emergence of life, period. The numbers juggled by the proponents of intelligent design are worthless.

Doesn't this cut both ways? If our ignorance is as vast as I've suggested, can Darwinians claim to know that life could have originated, and diversified, by the types of processes they currently recognize? The questions are rightly posed, for intelligent design can easily inspire a dogmatic overreaction. In its negative stance, the intelligent design movement identifies some unsolved problems for Darwinian evolutionary theory, and claims, illegitimately, that they are unsolvable. Until these problems are actually solved, however, it will be reasonable to wonder whether they can be disposed of without extending the resources of orthodox Darwinism.

So perhaps a full understanding of the history of life requires something else, besides the mechanism of natural selection? And, if something else, then why not intelligent design? Among contemporary evolutionists, a few look for alternative physicochemical mechanisms that might explain the "origins of order," or might yield understanding of the evolution of complex traits and structures.[72] The fans of intelligent design have failed lamentably to demonstrate that the unsolved problems of Darwinism are unsolvable. Yet, while we are seeking solutions, shouldn't we welcome their perspective as another pointer to research?

* * *

So far I have focused on the negative doctrine of intelligent design, the identification of unsolved evolutionary problems. We now have to consider the positive thesis, the claim that the phenomena to which Darwin's detractors point are produced by a process that deserves the label "intelligent." Two issues need to be considered. First, on what grounds should we apply the label? Second, what help can intelligent design provide in understanding the phenomena in question?

It's important to play by the rules. We mustn't personify Intelligence. Instead, according to the official doctrine, the acquiescence in a single tree of life, there are some evolutionary transitions—the original formation of life from nonlife, the emergence of the bacterial flagellum and the blood-clotting cascade—that can only be understood in terms of the action of a mechanism other than selection, that is, in terms of the operation of Intelligence.

As I've noted, there are some scientists who focus on kindred phenomena and try to find alternative physical mechanisms that would substitute for or complement natural selection.

They don't dignify their proposed mechanisms with the label "intelligent." So we must inquire what grounds might support this title.

Making any judgment about whether a mechanism is intelligent or not appears rather difficult until we have been told considerably more about the way in which that mechanism operates. Officially, of course, we aren't supposed to personify this mechanism, and it's hard to understand just what the attribution of Intelligence even means if we resist the personification. If something counts as intelligent, wouldn't it have psychological states and engage in psychological processes—and wouldn't anything like that be very like a person? Intelligent design-ers do not address such questions. All we learn from the full gamut of their literature, is that they conceive of Intelligence as whatever it is that produces the outcomes they identify as too complex to be attained through the operation of selection. The line of reasoning seems to be this: these phenomena, unattainable by selection, look designed or planned, and, as a result, the mechanism that produced them must be intelligent.

There is a fallacy here. Even without the slightest characterization of the mechanism, we're meant to infer one of its characteristics from the appearances of its products. One of Darwin's great achievements was to argue that you can have the appearance of design without a designer. Of course, even if they allow a limited role to natural selection, intelligent design-ers will contend that the kinds of complex organization to which they point couldn't have been produced by Darwin's surrogate for a designer—natural selection. Yet if we forget about natural selection, and ignore the controversies about what it can and cannot do, there are plenty of other instances in which striking order, pattern, and even beauty emerge from processes in which there is no planning, no design, but only the operation of blind and simple rules.

Consider standard celebrations of the free market. Frequently, these not only contend that large-scale order can emerge without planning, but also that attempts to plan, however sensitive and ingenious, would be bound to do far worse. The classic starting point for positions of this type is a famous passage of Adam Smith's. Speaking of an entrepreneur taking stock of the possibilities, Smith writes, "By preferring the support of domestic to that of foreign industry, he intends only his own security; and by directing that industry in such a manner as its produce may be of the greatest value, he intends only his own gain, and he is in this, as in many other cases, led by an invisible hand to promote an end which was no part of his intention. Nor is it always the worse for society that it is no part of it. By pursuing his own interest he frequently promotes that of the society more effectually than when he really intends to promote it."[73] Following Smith, generations of social scientists have argued, with different degrees of success in different cases to be sure, that the actions of myopic individuals busily pursuing their own concerns can generate a larger order that none of them had planned. At least some of these explanations are hard to resist.[74]

Or consider some contemporary studies of the ways in which the arrangements of petals in flowers, the banding patterns on seashells, the "designs" on the wings of butterflies and moths, and the coat patterns of mammals are generated in the development of those organisms—studies that give rise to strikingly beautiful computer graphics. In all these cases, researchers have been able to show how relatively simple—

unintelligent, unplanned—processes can be iterated to yield structures that look so intricate that one would naively take them to have been planned by an intelligent designer.[75]

These examples—and others can be found in the phenomena of chemical reactions that organize striking patterns, in the honeycombs of the beehive, and in the structures of snowflakes—reveal that we have to be careful in inferring the character of a causal process from the order we think we discern in its outcome.[76] It's simply a fallacy to suppose that because a particular structure or mechanism appears complex, then the causal agent that brought it about must be appropriately characterized as having "foreseen" or "planned" or "designed" the outcome. Even if intelligent design-ers were right in supposing that the phenomena they indicate couldn't have evolved by natural selection, only a more explicit identification of the causal mechanism that was at work could justify the conclusion that that mechanism is intelligent.

So, turning to the second question posed above, what help can intelligent design provide when we try to understand the difficulties it takes to beset Darwinism? How does it deal with the bacterial flagellum, for example?

If we take Behe at his word when he declares that he finds "the idea of common descent" to be "fairly convincing," and that he has "no particular reason to doubt it,"[77] then we should suppose that bacteria with flagella emerged from ancestors who lacked flagella. In line with the simple additive story he uses to make a history of natural selection appear implausible, he must suppose that the ancestors were missing a number of crucial proteins that the lucky descendants acquired, proteins that, once present, fit themselves together in the flagellum. If the intelligent design perspective is to help settle the unsolved problems of evolution, it would be good to have an alternative account that tells us how Intelligence facilitated the transition.

Unfortunately, the rest is silence. Neither in Behe's writings, nor in those of any other intelligent design-er, is there the slightest indication of how Intelligence performs the magic that poor, limited, natural selection cannot. On the face of it, there are just two basic possibilities. The first option is that Intelligence arranges the environment so that the intermediates—the apparently hapless organisms, cluttered with useless proteins—are protected against elimination under natural selection. (If we were unofficially inclined, we might say that the good Lord tempers the wind to the shorn bacterium.) The second option is that Intelligence provides for coordinated mutations to arise. If 20 genetic changes are needed, it brings about all of them at once. Or we can mix elements of both options and suppose that Intelligence introduces mutations in clumps—first ten, say, and then another ten, or first seven, then another seven, then six—and arranges protective environments for the intermediates. Of course, any story along these lines raises serious doubts. Just how does the coordination of the genetic changes or the modification of the environment work?

In presenting the possibilities in this way, I may seem to be forcing words into the mouths of the intelligent design-ers. Their core position, after all, is that at crucial moments in the history of life, descendants of some ancestors who lacked some trait (or organ or structure) came to possess the pertinent trait (organ, structure) by some causal process that is, unlike natural selection, intelligent. Why, then, do they have to talk about genes, mutations, and the need for protection against natural selection? The answer is that the traits in question are heritable—they are not introduced in each generation by some continued activity on the part of Intelligence, but emerge through the interactions of genes and environments. As in the case of the bacterial flagellum, there are underlying genes, and hence there have to be genetic changes in the passages from the ancestors to the descendants. If these changes occur over several generations, then, on the intelligent design-ers' own principles, there has to be protection against the tendency of natural selection to weed out the hapless intermediates. If they happen in one step, then, again by the favored principles, there must be coordinated mutations. Hence, even if the position would prefer to talk more vaguely of "novelties," it is committed to one of the options I have presented.

What intelligent design urgently needs if it's going to make any progress in understanding these transitions, in tackling the problems it claims to raise, is a set of coherent principles that identify the ways in which Intelligence is directed and what its powers and limitations are. If we lapse from the official story for a moment, we have to have some idea about what Intelligence "wants to achieve" and what kinds of things "it can do to work toward what it wants." What basis do we have to think that Intelligence aims to remedy the plight of the flagellumless bacteria, who can't evolve into bacteria-with-a-flagellum under natural selection? What basis is there to believe that Intelligence—or anything else, for that matter—can coordinate genetic changes or modify environments?

In fact, we need two distinct kinds of principles. First, there have to be principles that specify when Intelligence swings into action. Perhaps they will tell us that Intelligence operates when there are potentially advantageous complex traits that can't evolve by natural selection. Second, there must be principles that explain what Intelligence does when it acts. Perhaps these will identify the sorts of genetic changes Intelligence can arrange, or the ways in which it can inhibit the normal operation of selection.

It is already clear that these principles will be hard to state precisely. For, if Intelligence has been waiting in the wings throughout the history of life, seizing opportunities as they arise, we know that there are all sorts of things it hasn't done. Apparently Intelligence isn't directed toward eliminating the junk from genomes or removing vestigial structures like the whale's pelvis or generating radically new arrangements for mammalian forelimbs. It's possible, of course, that although directed toward these ends, Intelligence is simply unable to bring them about. So any satisfactory principles must differentiate between the bacterial flagellum, blood-clotting cascade, and similar places where Intelligence shows its prowess, and the accumulated junk, vestigial structures, and genetic blunders, where it remains in abeyance.

Drawing this distinction is even more difficult than I have made it appear. For there are really simple genetic problems with respect to which Intelligence seems to be impotent.

As I noted earlier, sickle-cell anemia has persisted because the gene that gives rise to the disease, when present in double dose, the *S* allele, confers an advantage when it appears in combination with the standard *A* allele, by providing resistance to malaria. This is the simple part of a classic evolutionary story, one routinely taught to those schoolchildren fortunate enough to learn something about evolution.

However, there is a twist, one not so widely known. In some African populations, there appears another form of the gene, the

C allele, usually only present at a low frequency.[78] The comparative rarity of the *C* allele is initially puzzling, for people who have two copies have the best of both worlds—they enjoy the resistance to malaria, and they aren't silent carriers of a deadly disease. Given its apparent advantages, why doesn't the *C* allele drive the others out? Why do inferior genes persist?

The sad answer is that natural selection tends to drive the *C* allele out of the populations in which it makes its cameo appearances. When it is relatively rare, its carriers tend to produce offspring with people who have the standard alleles—the *A* allele and the *S* allele—and, unfortunately, the combinations of *C* with *A* and *C* with *S* are inferior to both the more frequent genotypes (*AA* and *AS*). If only the *C* allele could reach a threshold frequency, then the chances of its bearers mating with one another and having children with the *CC* genotype would be sufficiently great to outweigh the defects of the *CA* and *CS* combinations—and, at that point, natural selection would eliminate the *A* and *S* alleles, producing a population with a genotype that would offer greater protection against disease.[79]

If Intelligence can handle the transition to the bacterial flagellum, then, according to the story on which Behe relies to discredit natural selection, it must be able to coordinate genetic changes or to provide environmental protection for otherwise vulnerable intermediates. Indeed, since there are, on Behe's hypothesis, a number of proteins that must be introduced to make the flagellum, it must be able to perform one or the other of these tricks on a rather grand scale. By contrast, the genetic problem with the *C* allele is trivial. It could be solved if Intelligence could protect a few organisms—those with the *CA* and *CS* genotypes—from the rigors of selection until the *C* allele had reached the requisite threshold frequency.

It could also be solved if Intelligence could arrange for some coordinated mutations, not a long series as with the bacteria, but just some extra *C* mutants. In fact, if Intelligence could simply produce one *C* allele in a human male and, simultaneously, a *C* allele in a human female, and then ensure that *C*-bearers mated with one another for a few generations, everything would work out beautifully. Apparently, however, Intelligence isn't up to these jobs. But if it can't perform these easy tasks, why should we think it can manage the transition to the bacterial flagellum?

Perhaps, however, the problem isn't with the power of Intelligence, but with its direction. Intelligent design-ers might suppose that their favored mechanism responds to the "needs" of the flagellumless bacteria, but not to those of human populations in which the *S* allele—and the debilitating disease it brings—is maintained by natural selection.[80] If we were to talk illicitly, we might wonder why Intelligence is more "worried about" the disadvantages that beset unflagellated bacteria than about the human beings who are, according to religious tradition, the foci of divine concern. Even if we play by the rules, however, it's appropriate to ask for the set of principles that govern the direction and power of Intelligence. Can intelligent design-ers explain how the problematic evolutionary transitions—those beyond the scope of natural selection—are accomplished, how Intelligence fails to overcome relatively small genetic barriers, and fails to clean up the repetitive DNA, purge the vestigial structures, and so on?

Why do intelligent design-ers ignore the basic problem of explaining the power and direction of the mechanism they invoke, a problem that strikes at the heart of their theory?

Apparently, their preferred perspective faces a multitude of currently unsolved puzzles about the scope and direction of Intelligence. Yet, unlike their counterparts in other scientific ventures, they are reluctant to suggest their strategies for seeking solutions. Their reticence provokes the charge that what they are doing is not science, but perhaps breaking their silence would be theologically unwise. Saying too much might disrupt the harmony between the sanitized version of intelligent design elaborated in the classroom during the week and the richer account delivered from the pulpit on Sunday. Moreover, saying anything that would genuinely respond to the puzzles might be saying too much.

Yet, I suspect many people would simply reject the terms in which I have posed the problem. Friends of intelligent design would prefer not to talk about evolutionary transitions at all. So, they might say, the complex structures are built from scratch. Intelligence is a creative force that replaces older types of organisms with new, individually designed species. Conceived in this way, intelligent design disavows Behe's acceptance of descent with modification, drawing bars across the tree of life to mark the places of radical discontinuity, of events of special creation. Intelligent design-ers stand forth as novelty creationists.

No canny opponent of Darwinism wants to say this too loudly, at least not when the discussion focuses on the scientific credentials of intelligent design. For, as we've seen, the retreat from Genesis leads to Lyell's alternatives, and a serious discussion of Lyell's alternatives dooms novelty creationism. Indeed, the discussion of the direction and power of Intelligence underscores the fundamental failure of novelty creationism, exposing a new kind of whimsy in the creative agency that produces flagellated bacteria and fails to produce human beings with functional hemoglobin as well as malarial resistance.

The real situation is that intelligent design-ers oscillate, officially rejecting novelty creationism when it seems strategic to do so, covertly adopting creationist ideas when there are hard questions about how evolutionary transitions are managed. Yet however they wriggle, they find no satisfactory positive doctrine, no set of principles about Intelligence that can adequately account for the phenomena. This is why readers hunt through their literature seeking fragments of positive theory in vain. They won't find it. Because to advance any such theory would expose the corpse of dead science.

* * *

Is my judgment too harsh? As I conceded above, problems about the origins of life on our planet are extremely difficult, and Darwinians have only been able to offer scraps of partial solutions. Perhaps this is one place at which intelligent design might offer a genuine alternative.

Yet the difficulty here is precisely the same as that exposed by the case of the bacterial flagellum. One might initially suppose that the residual problems of Darwinian efforts to explain the emergence of life from nonlife—the formation of proteins, the coordination of proteins and nucleic acids, the assembly of a package of hundreds, or thousands, of molecules in a primitive cell—could be cleared up by abandoning the complexities of biochemical research for the invocation of Intelligence. The illusion of an explanation can easily be generated, for it's natural to conceive Intelligence—illicitly—in the image of the divine Creator, whose hand reaches into the primeval soup (or slime?) and skillfully assembles the DNAs and the RNAs, the enzymes and the lipids. If you staunchly resist that temptation —as intelligent design-ers claim to think you should—then the story is much more austere. Natural selection cannot produce and assemble all these molecules. Intelligence can.

How, exactly? If there is anything more than an empty word in this doctrine, we need, as before, to know how Intelligence is directed and what its powers and limits are. Under what circumstances is Intelligence directed to produce complex structures? What kinds of complexities are its targets? (If we are to be careful and austere, we can't assume that Intelligence—like the God of the scriptures—wants to create life, and wants this kind of cellular life in particular.) What powers does it have in producing these complex structures? Until we have answers, there's no doctrine at all, and no genuine explanation has been given.

Moreover, any attempt to provide explanations is limited by problematic constraints, just as we saw in the case of the bacterial flagellum. To give sweeping powers to Intelligence would raise issues about the messiness of the history of life. Why does Intelligence not eliminate the accumulations of junk and structures that have lost their original functions? Why doesn't it create new species from scratch, without burdening them with the relics of the past? To give lesser powers would make it hard to understand how the extremely difficult task of the initial assembly of cells was carried out at all. To suppose that Intelligence is only directed toward giving life a kick-start would abandon all the arguments about the failures of natural selection throughout the major evolutionary transitions— and would commit Intelligence to a whimsical tolerance of bungled designs that is likely to be theologically troubling.

Intelligent design-ers sometimes protest that they do not need to answer these questions, claiming that the explanatory burdens for their proposals are rather different from those that attend Darwinism. In his promise to answer the toughest questions about intelligent design in his book, *The Design Revolution*, William Dembski maintains that intelligent design "is not a theory about a process but about *creative innovation*."[81] Yet, one might suppose, the results of creative innovation are frequently understood in terms of the intentions and powers of creators. Dembski sidesteps the point. "But intelligences are free. In the act of creation they violate expectations. They create as they choose to create. There's nothing that required Mozart to compose his Jupiter Symphony or Bell to invent the telephone or Shakespeare to write *King Lear*."[82] Fair enough—and it is reasonable to conclude, as Dembski does, that the innovators' contemporaries could not have predicted their innovations. The crucial issue, however, is whether we can understand the products, after the fact, as expressions of the purposes and the capacities of the creative agents. That appears to be precisely what musicologists and historians of technology and literary scholars sometimes attempt to do. By exactly the same token, if some past "innovation" in the history of life produced a bacterium with a flagellum, there are serious explanatory questions about how this living product represents the purposes and powers of Intelligence. If intelligent design is to earn respect as a genuinely scientific alternative, it cannot dodge such questions, but must take on the task of illuminating the events in the history of life after they have occurred.

The time has surely come to articulate intelligent design not as its savvy advertisers present it, but as the majority of those drawn to the position conceive of it. For them, matters are relatively simple. The unsolved problems of Darwinism in accounting for the origins of life provide grounds for believing an alternative. An all-powerful God, who delights in creation, brought life into being on our planet. Perhaps this was not quite done in the way that Genesis, read literally, would suggest. Perhaps life on earth does have a long history, but it has been guided throughout by the Creator. At times, God may even have intervened again, to create new life forms, and, when we find alleged evolutionary transitions that natural selection would seem unable to effect, we recognize the marks of these acts of creation.

One of David Hume's great insights was to ask what the character of the creation shows about the character of the Creator. Although that question might strike the devout as frivolous, or impertinent, it was very much in the spirit of Hume's own time, a time when reverent spirits investigated nature for traces of the divine purpose. With the explosion of our knowledge of how nature works, Hume's question has renewed force. Why did the Creator produce life by such lengthy and roundabout means? Why did He assemble primitive cells, and preserve elements of their structures in all subsequent organisms? Why has He permitted defects and junk to persist throughout his creation? Why is He unable—or unwilling—to solve the simplest genetic problems? Religious people reply that we cannot question—we are unable to understand the power and the purposes of God. That is a simple avowal of the point I have been making. If Intelligence is (unofficially) taken as a surrogate for God, we are unable to understand Its aims and capacities. No principles about its direction or about its powers can be stated.

Without such statement, however, intelligent design has no positive doctrine. Without positive doctrine, it provides no explanations. Without explanations it has no standing as a scientific alternative to Darwinism. And so the status claim made on its behalf is unwarranted. It is empty. Dead.

* * *

Fleeming Jenkin gave Darwin much trouble because, given what was then common understanding, the theoretical possibility of natural selection was suspect and the rate at which it could cause evolutionary change appeared far too slow. From the perspective of the 1860s, inheritance appeared to blend the traits of parents and the timescale for life on earth seemed confined to some hundreds of thousands of years. Late Victorian researchers did not know the things they thought they did, but they were entitled to worry that natural selection might not be the chief agent of evolutionary change.

Intelligent design-ers give contemporary Darwinians much trouble—but not because they pose intellectual challenges of the caliber of Fleeming Jenkin's. Their negative thesis consists of flawed attempts to show that currently unsolved problems are unsolvable without invoking their preferred alternative. Instead of arguing on the basis of established knowledge, they make unjustified assumptions about issues where we are—and know that we are—quite ignorant. Moreover, when it is scrutinized, their alternative turns out to be explanatorily bankrupt, hapless with respect to the very problems it raises as crucial difficulties for Darwinism. Indeed, the scientists who denounce intelligent design as not science are, I believe, responding to this explanatory bankruptcy.

I think we can now understand what is right and what is miscast in these denunciations. The temptation is to assume that there's a special type of specious doctrine, pseudoscience, marked out by some criterion from the genuine article, and that the people who espouse doctrines of this kind are pseudoscientists. I suggest that making the primary distinction between science and pseudoscience gets matters exactly backwards. There are people who assume the trappings of science, who go through the motions and talk the talk. They are pretenders, and thus appropriately seen as pseudoscientists. Pseudoscience isn't marked out by any clear criteria that distinguish it as a body of doctrine. It is simply what pseudoscientists do in their more-or-less-ingenious pretenses.

Justifying accusations of pretense is hard, requiring substantial study of the behavior of the accused. I hope to have said enough about the bobbings and weavings of intelligent design-ers, their rhetorical tactics, and their backing away from serious questions, to show why there might be a basis for the accusation. In the end, however, the label doesn't matter. Despite all the loud claims made by the resurrection men, they have failed to revive the discarded doctrines of the past.

My critique has taken intelligent design at face value, supposing that it attempts to provide a scientific alternative to Darwinian orthodoxy. Perhaps, however, there is an alternative interpretation, one that recognizes the intelligent design-ers as intending only to identify the limits of natural science.[83] On this reading, they would argue only that some evolutionary transitions cannot be understood in terms of the operation of natural selection—or indeed in terms of any other natural process—and they would modestly decline to advance any explanation of why such transitions have occurred. So construed, they would still be vulnerable to my scrutiny of their alleged proofs of impossibility, but they could evade my challenge to specify the principles under which Intelligence operates. They are simply not in that explanatory, that scientific, business. The envisaged retreat is unattractive. Not only would committed reticence make the claim that any natural explanation is impossible even less plausible, but it would also have the crippling disadvantage of forfeiting any claim to be doing science. Whether the resulting position would be even theologically satisfactory is a question we shall explore shortly.

Intelligent design-ers give Darwinians much trouble because their officially sanitized doctrine is presented with a wink.[84] That wink signals to the sincerely religious that there is a faith-friendly alternative to godless evolutionism, and so recruits to the cause people who have no part in any pretense and no idea that they are supporting an illusion. The zeal with which that cause is defended is the source of political—not intellectual—trouble for Darwinism. My next, and final, task is to try to understand the sources of this zeal.

Chapter Five

A MESS OF POTTAGE

For Wales? Why Richard, it profits a man nothing to give his soul for the whole world—But for Wales, Richard, for Wales!

Spoken by Thomas More, in Robert Bolt, *A Man For All Seasons*

When the Kansas School Board approved a characterization of science that would allow for the teaching of intelligent design in high-school biology, champions of Darwinism were quick to issue a stern warning. States that subordinate serious scientific education to religious concerns cannot expect to be attractive to businesses that are advancing, or applying, new forms of technology. In consequence, their citizens will lose opportunities for exciting and lucrative employment. More generally, American failure to honor established science will lead to a national decline in preparedness for the economic challenges of coming decades, as schoolchildren protected from disturbing ideas will no longer be able to compete in global markets. Darwinian commentators conjured the vision of the booboisie (in Mencken's famous phrase), burying their heads, ostrichlike, in rural idiocy, and thereby precipitating the decline of the West.

Many crucial, and controversial, assumptions are necessary to infer that the inclusion of intelligent design as part of science will lead to the inevitable impoverishment of scientific education and the demise of American (or Kansan) competitiveness. It is far from obvious that the ills of scientific education in the United States stem from squeamishness about teaching Darwin, or that the production of virtuosos in the scientific fields most relevant to economic growth will hinge on whether a large segment about evolution—unbalanced by any mention of intelligent design—figures in the curriculum. Even if these conclusions were correct, however, I doubt that stern warnings would strike those who want an inclusive definition of science, who want their children to hear about intelligent design, as penetrating to the heart of the matter. Quite reasonably, they would see the warnings themselves as another expression of the attitudes and values they reject.

In Robert Bolt's dramatization of the fall of Thomas More in *A Man For All Seasons* the protagonist is tried for his refusal to swear the oath acknowledging Henry VIII as the head of the English church. As the trial proceeds, More observes that his former protégé, Richard Rich, is wearing an impressive gold chain, and he asks Rich to explain its significance. When Rich replies that he has been promoted to a high post in the administration of Wales, More gently questions the choice, the venal bargain, he has made.

To use Mencken's vivid label, and to think of the worried Christian parents as the "booboisie," is a peremptory way of distancing and dismissing a serious perspective. Like Bolt's hero, the thoughtful people who want their children to learn about intelligent design view the warnings about jobs and the global economy as a crass failure to see what is genuinely important. An older story (Gen. 26, 31–3) might be more salient for them, the tale of the bargain between Jacob and Esau. Jacob is in the field, cultivating his crops, when Esau returns faint from the hunt. He asks his brother for food.

> And Jacob said, Sell me this day thy birthright.
> And Esau said, Behold, I am at the point to die: and what profit shall this birthright do to me?
> And Jacob said, Swear to me this day; and he sware unto him: and he sold his birthright unto Jacob.

To trade your soul for Wales, to let Darwinism unchecked into the schools, these are equivalent, equivalently myopic, bargains, exchanges in which you give up your birthright for a bowl of lentils—for a mess of pottage.

In fact, it is worse. The birthright, or the soul, you give up is not your own, but that of your child. Instead of doing your duty for your sons and daughters, instead of steering them in the faith and helping them to God, you open hatches through which they may fall, jeopardizing their chances of salvation in hopes that they may obtain some trivial mundane reward. I know from my own case how poignant a child's loss of faith can seem to a religious parent. During my teens, although I continued to sing in church choirs, it became evident to my family that I no longer believed in the Christian doctrines in which I had been brought up—I made no secret of my abstinence from the sacraments, for example. For my mother, the education I had received, once seen as wonderful, beyond any dreams she could have had for me, now appeared as terrible, the source of a turn in my life that had deprived me of the single most precious possession any person can have. The worldly successes my education promised me were completely irrelevant, utterly incapable of assuaging her sorrow.

I grieved for her grief, but could not—cannot—see things as she did. Yet even though I do not think there is any birthright I have traded for a mess of pottage, any sense in which I have given my soul for Wales, I believe that adequate understanding of the resistance to Darwinism must recognize the perspective that takes acceptance of orthodox biological education as a terrible bargain. Enough has been said in previous chapters about the fallacies purveyed by the intelligent design-ers. My concern now is with the honest and worried people who accept the advertisements for intelligent design with a sense of liberation, and with the values they hope to preserve.

* * *

The simplest way to address worries about the effects of teaching evolution is that pioneered in Westminster Abbey over a century ago. From the eulogies at Darwin's funeral to eloquent contemporary presentations,[85] religious people have argued that the opposition between Darwinism and faith is only apparent. If you can have God and Darwin too, then the concerns of those who support intelligent design can be met without grasping at the illusions the design-ers concoct. From this perspective, the controversy persists because of two mistakes: one that fails to see how the evolution of life on earth, caused by natural selection, might elaborate the plan of a Creator who set things up to proceed in this way and who leaves natural processes to run their course, and another that disguises currently unsolved difficulties with Darwinism as unsolvable, thereby creating space for a much-touted, but ultimately empty, alternative. A nonproblem is created, and a spurious solution is then offered.

According to this diagnosis, the honest supporters are doubly deceived. They are bombarded with the admonishment that they must choose between God and Darwin. Then they are told that, despite the widespread scientific support for Darwin, there's another point of view, defensible on scientific grounds, and unfairly derided by the academic establishment. Those who say these things appear more trustworthy than the remote Brahmins who pooh-pooh opposition to evolution. The audience has little reason to think critically about the story they are told. Many of them find the scientific details difficult, and, when they contrast the reassurances of their trusted counselors with the haughty dismissals of a secular orthodoxy, their sense of alienation from science deepens.

I agree with some parts of this account. With any major piece of science, it is possible to identify unsolved problems, and to conjure up a "case for balance," a case that would require significant work and attention to expose it for the charade it is. Present the "case for balance" in evolutionary theory to people who are already worried about the impact of Darwinian ideas on their children, people who lack the tools to identify its chicanery, people who don't have the motivation to probe it as they would other heterodox claims, and it's highly likely that you'll succeed in rallying them to the cause. Where I demur, however, is in the thought that the worries about Darwinism are themselves unfounded, that the supporters of intelligent design have misguidedly erected a non-existent opposition between Darwinism and the religious doctrines central to their faith.[86]

From the late nineteenth century on, religious people who have thought hard about the Darwinian view of the history of life have found it deeply troubling. George John Romanes, author of books on religion and works of science, found Darwin's vision agonizing. It seemed to him that the universe had "lost its soul of loveliness."[87] In his groundbreaking *The Varieties of Religious Experience*, William James articulated more extensively this sense of loss, offering an arresting image. "For naturalism, fed on recent cosmological speculations, mankind is in a position similar to that of a set of people living on a frozen lake, surrounded by cliffs over which there is no escape, yet knowing that little by little the ice is melting, and the inevitable day drawing near when the last film of it will disappear, and to be drowned ignominiously will be the human creature's portion. The merrier the skating, the warmer and more sparkling the sun by day, and the ruddier the bonfires at night, the more poignant the sadness with which one must take in the meaning of the total situation."[88] Given this picture of life as early twentieth-century science seems to depict it, James can only view cheerfulness, or even the absence of despair, as based on false optimism, on failure to face reality. It is hardly surprising that he sees religious impulses as cries, triggered by the need for something different or for something more—"Here is the real core of the religious problem: Help! Help!"[89]

Perhaps this is overwrought, even neurotic? I don't think so. Romanes and James, like the evangelical Christians who rally behind intelligent design today, appreciate that Darwinism is subversive. They recognize that the Darwinian picture of life is at odds with a particular kind of religion, providentialist religion, as I shall call it. A large number of Christians, not merely those who maintain that virtually all of the Bible must be read literally, are providentialists. For they believe that the universe has been created by a Being who has a great design, a Being who cares for his creatures, who observes the fall of every sparrow and who is especially concerned with humanity.[90] Yet the story of a wise and loving Creator who has planned life on earth, letting it unfold over four billion years by the processes envisaged in evolutionary theory, is hard to sustain when you think about the details.

Many people have been troubled by human suffering, and that of other sentient creatures, and have wondered how those pains are compatible with the designs of an all-powerful and loving God. Darwin's account of the history of life greatly enlarges the scale on which suffering takes place. Through millions of years, billions of animals experience vast amounts of pain, supposedly so that, after an enormous number of extinctions of entire species, on the tip of one twig of the evolutionary tree, there may emerge a species with the special properties that make us able to worship the Creator. Even though there may be some qualitative difference between human pain and the pain of other animals, deriving perhaps from our ability to understand what is happening to us and to represent the terrible consequences, it is plain to anyone who has ever seen an animal ensnared or a fish writhe on a hook, that we are not the only organisms who suffer. Moreover, animal suffering isn't incidental to the unfolding of life, but integral to it. Natural selection is founded on strenuous competition, and although the race isn't always to the ruthless, there are plenty of occasions on which it does produce "nature red in tooth and claw" (in Tennyson's pre-Darwinian phrase). Our conception of a providential Creator must suppose that He has constructed a shaggy-dog story, a history of life that consists of a three-billion-year curtain-raiser to the main event, in which millions of sentient beings suffer, often acutely, and that the suffering is not a by-product but constitutive of the script the Creator has chosen to write.

To contend that species have been individually created with the vestiges of their predecessors, with the junk that accumulates in the history of life, is to suppose that Intelligence—or the Creator—operates by whimsy. The trouble is that the charge doesn't go away when the action of the Creator is made more remote. For a history of life dominated by natural selection is extremely hard to understand in providentialist terms. Mutations arise without any direction toward the needs of organisms—and the vast majority of them turn out to be highly damaging. The environments that set new challenges for organic adaptation succeed one another by processes largely independent of the activities and requirements of the living things that inhabit them. Even if the succession of environments on earth has some hidden plan, Darwinism denies that the variations that enable organisms to adapt and to cope are directed by those environments. Evolutionary arms races abound. If prey animals are lucky enough to acquire a favorable variation, then some predators will starve. If the predators are the fortunate ones, then more of the prey will die messy and agonizing deaths.[91] There is nothing kindly or providential about any of this, and it seems breathtakingly wasteful and inefficient. Indeed, if we imagine a human observer presiding over a miniaturized version of the whole show, peering down on his "creation," it is extremely hard to equip the face with a kindly expression.

Toward the end of the *Origin*, Darwin points to some striking, and disturbing, phenomena, which on the Lyellian alternative appear to signal the whimsy (or, perhaps, callousness) of the Creator. He suggests that his account, unlike the explanation in terms of special creation, should remove many surprises at the character of the living world. "We need not marvel at the sting of the bee causing the bee's own death; at drones being produced in such vast numbers for one single act, and then being slaughtered by their sterile sisters; at the astonishing waste of pollen by our fir-trees; at the instinctive hatred of the queen bee for her own fertile daughters; at ichneumonidae feeding within the live bodies of caterpillars; and at other such cases."[92] The last example is well chosen, for the behavior of the ichneumonidae—parasitic wasps—is particularly unpleasant. The wasps lay their eggs in a living caterpillar, paralyzing the motor nerves (but not the sensory nerves) so that the caterpillar cannot move or reject its new lodgers. As the eggs hatch, and the larvae grow, they eat their way out of their host.

Darwin presents his catalogue of surprisingly nasty aspects of nature to argue that taking these arrangements to result from acts of separate creation implies an extraordinary degree of whimsy on the part of the Creator. He doesn't make explicit an implication that disturbs many of his most sensitive readers—those for whom his universe has lost "its soul of loveliness"—the fact that matters are little better if the Creator's activity is more distant. The mess, the inefficiency, the waste and the suffering are effects of natural processes, so that they shouldn't be seen as directly planned and introduced. But the Creator has still chosen to use those processes to unfold the history of life. The general inefficiency of the processes, the extreme length of time, the haphazard sequence of environments, the undirected variations, the cruel competition through which selection so frequently works, is all foreseen. And the individual nastinesses to which Darwin points are expected outcomes of deploying these sorts of processes. If we search the creation for clues to the character of the Creator, a judgment of whimsy is a relatively kind one. For we easily might take life as it has been generated on our planet as the handiwork of a bungling, or a chillingly indifferent, god.

Sober consideration of the history of life is bound to generate just the questions that fueled the nascent skepticism of the eighteenth century, doubts directed toward providentialism. For Voltaire and for Hume, the ancient problem of the existence of evil in a world designed by a powerful and benevolent deity arose with new force. In his *Dialogues Concerning Natural Religion*, Hume placed a series of questions in the mouth of one of his characters. Are the evils unforeseen? Or is it that they are foreseen and the deity has no power to remove them? Or should we suppose that they are foreseen and recognized as removable but that the Creator simply chooses not to do so? The general lines of theological answer, well known to Hume, and repeated by many apologists since, are that the evils we perceive are merely local, and that they make an essential contribution to a larger good. (The apologies are principally the work of theologians and academic philosophers. Many devout people find them wrongheaded and insensitive, and would reject the entire idea of attempting to fathom the divine plan.)

Ambitious providentialists try to say what this greater good is, and why pain and suffering are necessary for it to exist. Virtues require adversity. Courage cannot exist without the threat of danger. Generosity is only possible if there is also want. Moreover, for genuine virtue to be present, people must be able to act freely. That means that there must be the possibility of sin. Hence we shouldn't be surprised if the powerful and loving Creator has brought about a world in which there is pain and suffering, some of it produced independently of human beings, some of it resulting from free human actions. If the most important good—the existence of people who freely act virtuously—is to be present in the Creation, then the dark and cruel aspects also must be included.

There is plenty to challenge here, and complex philosophical debates have raged around such questions as whether God could have created people who always freely chose to do the good. I shall be content with two simpler, and, I believe, more disturbing points. The first of these stems directly from the Darwinian picture of the history of life. When you consider the millions of years in which sentient creatures have suffered, the uncounted number of extended and agonizing deaths, it simply rings hollow to suppose that all this is needed so that, at the very tail end of history, our species can manifest the allegedly transcendent good of free and virtuous action. There is every reason to think that alternative processes for unfolding the history of life could have eliminated much of the agony, that the goal could have been achieved without so long and bloody a prelude.

The second point is that the providentialist's doctrine that humans and nonhuman animals suffer in the interests of achieving some greater good must be reconcilable with the assumption of divine justice. You cannot defend torturing a few individuals who are known to be innocent on the grounds that setting some examples will contribute to a safer society. By the same token, a just Creator cannot consign vast numbers of its creatures to pain and suffering because this will promote some broader good. Divine justice requires that the animals who suffer are compensated, that the suffering isn't simply instrumental to the wonders of creation but redeemed for them. Dostoevsky's Ivan Karamazov presents the fundamental point. "It's not worth the tears of that one tortured child who beat itself on the breast with its little fist and prayed in its stinking outhouse, with its unexpiated tears to 'dear, kind God'! It's not worth it, because those tears are unatoned for. They must be atoned for, or there can be no harmony."[93] How, then, are the agonies, especially the agonies of the innocent, atoned for?

The providentialist may seem to have an obvious reply. Atonement comes in union with God, perhaps in an afterlife. Further reflection makes it evident that this only raises new difficulties.[94] If the suffering of the child, or of the holocaust victim, is genuinely outweighed by some greater good, received in an afterlife, we ought to ask if that suffering is necessary for that good to be received. If it is, then others who have been denied the suffering will turn out to have been shortchanged. They will not have experienced something that is necessary for the attainment of the greater good. Because such acute suffering is needed for salvation in the afterlife, those who do not suffer similarly cannot be saved. If, however, salvation is possible without the suffering, then the agonies are unatoned for. They weren't needed for the glorious reward. Finally, if it is suggested that there are two kinds of people, those who need to suffer to win eternal salvation and those who do not, we need to know just how the holocaust victims are different from those whose lives are free from comparable violations—and just why the divine plan demands the existence of some who can only attain heaven through extreme suffering.

I've been exploring ambitious providentialism, which attempts to respond to the existence of evil, pain, and suffering by explaining why they are necessary for the greater good of the Creation. Providential religion could be humbler, admitting that the reasons of God are beyond human understanding. It could concede that, when we look at the history of life, seeking clues to the character of its divine Creator, we're likely to be confused and misled, but that is because we are finite creatures, incapable of appreciating God's greatness or His purposes.

Yet this strategy of retreat comes at a price. First, it involves just that suspension of curiosity that moved Darwin to an atypically scathing critique. If you are prepared to treat the divine plan as ultimately mysterious and incomprehensible, then why introduce that thought just here? Why not go further? You might declare that the appearances of common descent are deceptive, that species have been newly created with the vestiges of formerly useful organs and structures, with the masses of genomic junk, and that the Creator has His own unfathomable reasons for doing this. You might even insist that the earth has been made with the appearance of great age, that the order of the fossils in the rocks and the radioactive residues are products of a recent Creation, that in all these instances the intentions of the Creator in mimicking a Darwinian world are beyond human understanding. Wherever it occurs, in defending a beneficent Creator against the evidence of whimsy or indifference, in advocating novelty creationism, or in resurrecting Genesis, the appeal to the incomprehensibility of the Deity faces the same objection. "It makes the works of God a mere mockery and deception; I would almost as soon believe with the old cosmogonists, that fossil shells had never lived, but had been created in stone so as to mock the shells now living on the sea-shore."[95] The appeal to "mystery" is always available—and always an abdication of the spirit of inquiry. For those who would reconcile God and Darwin, it's hardly an acceptable resting place.

Hume challenged his providentialist contemporaries by asking them to consider what character they would ascribe to the deity if they set aside their preconceptions and simply used the observed phenomena of life on our planet as the basis for their inference.[96] It might appear that the challenge is unfair, that there are occasions on which we suppose that appearances are deceptive, believe that what seems a natural conclusion from the observed phenomena should not be drawn, think that there is an—unknown—explanation for the discrepancy between the "obvious implication" and what we ought to accept. Not all unanswered questions are unanswerable. It would be entirely unreasonable for me to protest that there is no way to fill in the blanks in my unfinished crossword, or for the community of scientists to assert that there is no answer to some large question—the problem of protein folding, say—that currently baffles them. In some instances, we would properly assume that there is a solution to a problem, even though we recognize quite clearly that we shall never be able to provide it. There are vast numbers of questions of human history about which we'll always remain ignorant. At this point, however, a deeper problem emerges. For on the occasions on which we are justified in thinking that there is an explanation, currently or even permanently unbeknownst to us, we have background knowledge to which we can appeal. Although we cannot say what route Caesar followed on the Ides of March, the information we have provides grounds for thinking that he followed some definite course to the Capitol. If the providentialist is to turn back Hume's challenge—or the Darwinian extension of it—then it must be because there are antecedent grounds for supposing that the providential Creator exists. Were that not so, then there would be no basis for supposing that the waste, the suffering, and the inefficiency should not be taken at face value.

Troubling questions now arise. Why should anyone think there must be a providential Creator behind the apparent evils of the world, a God whose purposes we cannot fathom? On what do providentialists rely when they maintain that there must be some unknown order behind the messiness of life?

So the conflict between Darwin and providential religion leads inexorably into a broader battle. It pitches us into what is often (but wrongly) viewed as a war between reason and religion generally, one that erupted in the eighteenth century and that has intensified ever since. Darwinism is entangled with what I'll call the "enlightenment case against supernaturalism." Evolutionary ideas form a separable part of the case, as well as amplifying other themes within it. It is wrong to give Darwin complete credit as the "anatomist of unbelief." But it would also be wrong to pretend that his ideas are not important to the "delineation of doubt." I shall try to explain below why he is so prominent a figure in the conflict, why he serves evangelical Christians as the bogeyman.

* * *

The enlightenment case began with attacks on providentialism, but the vast majority of the world's religions have not been committed to a wise and powerful Creator with a great, if unfathomable, plan. An inclusive pantheon would contain many gods—and spirits, and ancestors—who have little interest in human or animal welfare, some of whom can be placated in various ways, most of whom have to be acknowledged as sources of power. An even more capacious collection of religious entities would include impersonal powers, forces like the Mana of some Polynesian or Melanesian religions, with which it is important to align oneself. It is not easy to identify what distinguishes these objects of religious concern, these gods and ancestors, spirits and forces, except to say, vaguely, that they are very different from the normal things with which human beings deal, that they are not perceptible except under very special circumstances, that they are somehow "supernatural" or "transcendent."

Religion is itself an extraordinarily diverse and multifaceted phenomenon, emerging in different forms in different societies, and even assuming new identities in a changing historical and social context. At different times, and in different locations, the major religions of the world, Judaism and Islam, Buddhism and Hinduism, as well as Christianity, have all embraced very different conceptions of the religious life. Because of this, an enlightenment case against religion is likely to fail. Religious traditions can evolve, adapting themselves to the arguments presented so that the skeptics' attempts to define "religion" once and for all are portrayed as limited and crude. Instead, I suggest, we should recognize an enlightenment case against a common strand within religious traditions, against supernaturalism.

Providentialist religion, as we have seen, supposes that there is an unknown—even unknowable—explanation for the mess of life, and, in doing so, relies on claims to know that there is a wise and benevolent Creator. The enlightenment case scrutinizes this claim to knowledge, and does so by opposing all alleged knowledge of supernatural (or transcendent) entities. It recognizes that many versions of the world's religions are committed to the existence of supernatural beings and to the truth of particular stories about these beings. Most religions that have existed thus far have been supernaturalist, that is, they have relied on oral traditions or canons of scripture that describe the characteristics and actions of supernatural entities, usually although not always supernatural persons, and acquiescing in the religion frequently requires belief that many of these descriptions are literally true. Devout Christians have typically believed that Jesus was once literally raised from the dead. Devout Jews have often supposed that God literally made a covenant with Abraham and, later, with the chosen people. Devout Muslims routinely think that the angel Gabriel spoke the exact words that the Prophet heard and recited, the words recorded in the Qur'an. Australian aborigines believe that important events occurred during what they call the "Dreamtime." Devotees of African religions make strong claims about the enduring presence of ancestors, and so on. The enlightenment case begins with an assault on the doctrines that are presupposed in providentialist Christianity, but it proceeds to attack all versions of supernaturalist religion. Indeed, as we shall discover, part of its strategy for undermining any particular religion involves a negative attitude toward supernaturalism in general.

Despite its scope, this is not a war against all religion. For there are other kinds of religion, "spiritual religions," as I shall call them, that don't require the literal truth of any doctrines about supernatural beings. Some professing Christians and professing Jews have heard the dispatches from the enlightenment front, and responded by abandoning commitment to the literal truth of virtually all the sentences in their respective Bibles. The possibility of spiritual religion will occupy us later. The next step in an investigation of the tangled relationship between Darwinism and religion must be a quick review of the enlightenment case against supernaturalism.

The enlightenment case against supernaturalism begins by asking for the grounds on which the devout might become confident that there must be some explanation for the pains and sufferings of sentient beings, for the waste and inefficiency of the history of life and the operation of natural selection. Providentialist Christians reply that they accept a body of background doctrine, which tells them of a powerful, wise, and benevolent Creator. They endorse this doctrine because they believe in the literal truth of certain statements in the Christian Bible.[97] They respect the authority of a particular church, or denomination, or tradition, and rely on the original revelation of divine truth, embodied in the sacred scriptures and unfolded by the learned in each generation.

The enlightenment case subjects this idea to intense examination. Can the long and intricate process that leads from some original event—a supposed revelation—through the formation of the texts, their dissemination and interpretation, provide any real basis for firm belief that the disorder and messiness of life is only apparent? Or are religious believers like children, deeply committed to a story inculcated by fond parents, children who declare that there must be some explanation that allows reindeer to fly and presents to be universally dispensed on Christmas Eve—even though no such explanation comes to mind? (They are firmly convinced, after all, that there is a Santa Claus.) From the eighteenth century to the present, dedicated scholars have probed the scriptures, tried to understand the circumstances of their composition, scoured the historical record to account for their acceptance as canonical, and have elaborated sociological explanations of the careers of some major religions. Their investigations suggest that the texts and traditions cannot support the confidence of the faithful—and it is hardly surprising that religious people usually know far less about these scholarly studies than they do about Darwinian evolution.

The canonical Christian texts are the Gospels and the letters attributed to Paul. Literary analysis and historical studies have established, as firmly as anything in ancient history is ever established, that not all the letters the New Testament assigns to Paul were written by the same person, but that the genuinely Pauline documents are the earliest part of the Christian canon, written about twenty years after the Crucifixion by a man who had never had any intimate association with Jesus and whose convictions had been altered by a critical event. The four Gospels, written about two decades later still, are incompatible with one another on many points of detail. Jesus does similar things and tells similar stories, but in different locations, to different audiences, or in a different temporal order. There are striking differences about the events after Jesus' death. The original version of the earliest Gospel—that of Mark—ends with an empty tomb; the brief description of an appearance of the risen Christ comes only in a later addendum to the text. Both Matthew and Luke provide much more elaborate stories of Jesus' appearances to the disciples, but the locations, and people involved, are remarkably different. John, the latest of the Gospels, provides the most extensive narrative, giving details of the appearances of the risen Christ that are at variance with those of all the others—and only John picks out an individual disciple, Thomas, who refuses to believe in the resurrection without seeing and touching. Not all these conflicting reports can be literally true.

There are similar difficulties with the beginnings of Jesus' life. Mark provides no details, but begins with the adult Jesus coming to the Jordan to request baptism by John the Baptist. Both Matthew and Luke, however, tell elaborate stories, and one of Matthew's formulations provides a clue about how they proceeded. A distinctive feature of Matthew's Gospel is his interest in connecting events in Jesus' life with Old Testament prophecies about the Messiah. (Overall, Matthew is most concerned with linking the nascent Jesus movement of his time —roughly 80 CE—to Jewish laws and traditions.) He turns frequently to Isaiah, using a standard Greek version of the text (the Septuagint), formulating a prophecy in language most Christians know well. "Behold! A Virgin shall conceive, and bear a Son. And shall call his name, Immanuel." The original Hebrew, however, is less biologically shocking, announcing only that a young woman shall conceive. The pre-Christian translators who crafted the Septuagint chose the Greek word *parthenos* (virgin), and thereby unwittingly inaugurated a piece of Christian theology, the Virgin birth—or, as I would prefer to put it, they led Matthew to create a myth.

The mythical character of these stories becomes ever more apparent when you compare the narratives offered by Matthew and Luke. Luke's moving version of the events surrounding the birth of Jesus requires Joseph and Mary to travel from Nazareth to Bethlehem. Matthew has no need of any such journey, since he locates them in Bethlehem all along. Matthew has wise men, but no shepherds. Luke has shepherds but no wise men. (The Christianity with which I grew up solved the problem in the obvious way by combining everything.) There are some serious difficulties in reconciling the dates. Herod, a main character in Matthew's story, died about ten years before the appointment of the Roman official whom Luke takes to have been an administrator in the region at the time of the nativity. But I want to focus on a different detail.

Because he wants the birth of Jesus to fulfill the prophecy that the Messiah would be born in Bethlehem, Luke has to explain why Joseph and the pregnant Mary made a journey from Nazareth. His solution is offered in a beautiful passage, one that rings out in churches each Christmas (Luke 2:1–4). "And it came to pass in those days, that there went out a decree from Caesar Augustus, that all the world should be taxed. And this taxing was first made when Cyrenius was governor of Syria. And all went to be taxed, every one into his own city. And Joseph also went up from Galilee, out of the city of Nazareth, into Judea, unto the city of David, which is called Bethlehem, because he was of the house and lineage of David."[98] The overwhelming evidence is that this is complete fiction.

Not only are there no records of a census or a general taxation at this time, but, even if there had been one, this is surely not the way in which it would have been conducted. We know something about Roman attitudes toward the religious lore and ethnic traditions of the Jews—at best, they saw them as barbaric enthusiasms. We also know something about the ways in which Romans obtained population counts and how they levied taxes. Instead of moving the people about, they quite sensibly dispatched their own trusted officials. Luke invites us to think of Cyrenius as having done something quite mad. In the interests of administering some kind of census or taxation, he encourages a mass migration to bring people to the places with which particular ancestors are associated, ancestors whose importance is fixed by Jewish culture.

As a last example, consider the depiction of Pilate offered by the Gospels. The familiar story of the Roman official offering to release Jesus and encountering a baying Jewish mob has played a significant role in Christian anti-Semitism. The action Pilate contemplates, releasing a prisoner for a local religious festival, is quite unprecedented in the Roman administration of Judea, or of any other province with indigenous zealotry. It is also incompatible with what we know of the man, some of whose repressive actions are documented, who was, apparently, recalled because of protests against his harsh treatment of the Jews. What accounts for the portrait of a sympathetic figure, so different from the indications we have from other sources?

The canonical Gospels were written as the expression of a broader Hellenistic Jesus movement, after it was clear that that movement was unlikely to flourish as a reform of Jewish religion,[99] and after the Roman grip on the eastern Mediterranean had tightened. The evangelist who first recorded the story, Mark, chose a strategy of appeasing the Romans and making scapegoats of the Jews. His choice was politically adept, and probably helped the movement appeal to non-Jews in a world dominated by Rome. Yet I concur with the judgment of the Jesus Seminar, a group of theologians that has brought the most developed textual scholarship to bear on the accuracy of the Gospel narratives. "That scene, although the product of Mark's vivid imagination, has wrought untold and untellable tragedy in the history of the relations of Christians to Jews. There is no black deep enough to symbolize the black mark this fiction has etched in Christian history."[100]

Only in the second century of our era was the Christian canon assembled. The New Testament as we have it is surely a compromise, designed to satisfy groups of believers with different favored texts and different oral traditions. Some followers of Jesus, however, were left out. Among them were those who treasured different "Gospels"—the Gospel of Mary, the Secret Book of James, and, perhaps most interestingly, the Gospel of Thomas.[101] This last text, discovered in 1945, overlaps with Matthew and Luke in reporting many of the familiar sayings and parables of Jesus, but also contains others that are startling.

> The disciples said to Jesus, "We know that you are going to leave us. Who will be our leader?"
> Jesus said to them, "No matter where you are, you are to go to James the Just, for whose sake heaven and earth came into being."[102]

Thomas, for whom the Gospel is named, has private conversations with Jesus, in which he is told secrets that cannot be revealed, insights that the other disciples cannot understand. Perhaps this is why the Gospel of John is so keen to debunk Thomas' credibility, to portray him not as the privileged recipient of higher truths, but as "Doubting Thomas."[103]

The documents Christians take to be canonical were chosen as the result of political struggles among many nascent Jesus movements, in which efforts to incorporate the ideas of an itinerant teacher within the framework of Judaism lost out to a more cosmopolitan vision favored by the Rome-oriented Paul. Within that cosmopolitan conception there were also variations, some of which were included in the compromise we have, others of which were eliminated as heretical. Out of this has come a collection of inconsistent documents, many of whose parts are evidently fictitious. How can reliance on this canon provide grounds for thinking that, despite all appearances, life has been planned by a powerful and benevolent deity?

The effects of scholarly study of Christian scripture have long been evident to those who have taught seminary students. In a famous letter of resignation, Julius Wellhausen, one of the great interpreters of the scriptures, acknowledged the effects of his discoveries. "I became a theologian because I was interested in the scientific treatment of the Bible; it has only gradually dawned upon me that a professor of theology likewise has the practical task of preparing students for service in the Evangelical Church, and that I was not fulfilling this practical task, but rather, in spite of all reserve on my part, was incapacitating my hearers for their office."[104] With reason, many evangelical sects keep the news from the scholarly front from the faithful. Often, the tactic chosen consists in flat assertions that deny the reasoned judgments of scholars about historical or linguistic matters. The *King James Study Bible* (designed, as the note to the reader explains, to provide a reliable guide for "conservative Christians") responds to the lengthy discussions about the priority of Mark and about the existence of an additional source from which both Matthew and Luke drew with a single sentence: "There is still very strong reason to hold to the priority of Matthew as the first gospel account of the life of Christ."[105] The faithful are given neither an extended account of what the scholarly consensus is, nor of what the "very strong" reasons are for rejecting it.[106]

* * *

Although I have been concentrating on Christianity, and on the difficulties involved in accepting the claims made by canonical Christian texts as literally true, similar points apply to other religious traditions. The sources of difficulty are typically the same. Extraordinary events are supposed to have taken place in the more-or-less distant past, to have been recorded in writing or passed down in oral recitation, so that people who live today should believe in the literal truth of the cherished stories. When the processes through which these stories have come down to us are examined, there are often grounds for doubt about the marvelous events that initiated the process, often internal contradictions in the variant versions, often signs of political struggles in formulating orthodoxy. This, however, is merely the beginning of trouble for an uncritical reliance on texts and traditions. As understanding of the diversity of the world's religions increases, it's hard for believers to avoid viewing themselves as participants in one line of religious teaching among many. You profess your faith on the authority of the tradition in which you stand, but you also have to recognize that others, people who believe very different, incompatible things, would defend their beliefs in the same fashion. By what right can you maintain that your tradition is the right one, that its deliverances are privileged?

For all their doctrinal disagreements, Muslims, Jews, and Christians agree on many things. If, however, you had been acculturated within one of the aboriginal traditions of Australia, or within a society in central Africa, or among the Inuit, you would accept, on the basis of cultural authority, radically different ideas. You would believe in the literal truth of stories about the spirits of ancestors and about their presence in sacred places, and you would believe these things as firmly as Christians believe in the resurrection, or Jews in God's covenant, or Muslims in the revelations to the Prophet. Victorian explorers, confident in the superiority of their race, their culture, and their particular version of Christianity, collected stories from the dark continents they visited, labeling them as primitive superstitions. They failed to observe that the cultural processes that generated the exotic beliefs they recorded were exactly the same as those that lay behind their own religious doctrines—in all instances, there is a long sequence of generations along which stories of extraordinary happenings are passed on as the lore of a social group. To what can a believer within any particular tradition—Christianity, say—point to show why it is uniquely right, and its rivals wrong?

The trouble with supernaturalism is that it comes in so many incompatible forms, all of which are grounded in just the same way. To label someone else's cultural history as "primitive" or "superstition" (or as both) is easy, until you realize that your basis for believing in the literal truth of the wonderful stories of your own tradition is completely analogous to the grounds of the supposedly unenlightened. There are no marks by which one of these many inconsistent conceptions of the supernatural can be distinguished from the others. Instead, we have a condition of perfect symmetry.

Perhaps the symmetry can be accepted. Perhaps all these traditions are fundamentally correct, and we should focus on the core doctrine on which all agree. Yet as comparative studies disclose more and more differences among religions, some polytheistic, some monotheistic, and some without any conception of a personal deity, the more attenuated any such "core doctrine" becomes. Hence you arrive at frustratingly vague definitions of "religion" that appeal to some "acceptance of the transcendent." Even if it were supposed that the traditions were right about the doctrines they hold in common, the specific stories—the resurrection, the covenant, the divinely inspired recitation—would have to be abandoned.

It would be necessary to move beyond supernatural religion to spiritual religion.

Social and historical studies of the growth and spread of major religions reinforce this point. Although we lack direct evidence for many religious traditions—including all those that flourished before the invention of writing—it is possible to recognize the features that have fueled the rise of successful modern sects, and to explore some historical cases. A crude hypothesis based on what evidence we have suggests that religions spread within societies when they offer members of the societies something they want. They spread across societies when they encourage social cohesion, and when they enable a society to deal successfully with its neighbors. The details are likely to differ from case to case, and a blanket claim of this sort is only the prelude to serious history. The principal point, however, is that religious doctrines don't have to be true to be successful. Truth, like Mae West's goodness, may have nothing to do with it.

Why did Christianity succeed in the Greco-Roman world? Statistics suggest that upper-middle-class pagan women were relatively more attracted by the religion—perhaps because they perceived the lives of their Christian counterparts as better than their own (that Christian husbands were more faithful and less abusive).[107] An intriguing conjecture proposes that, in an urban world marked by filth and recurrent outbreaks of plague, the Christian injunction to comfort the sick would have raised survival rates in times of epidemic, simply because of the beneficial effects of giving water and other forms of basic care. Outsiders would have seen that Christians recovered more frequently, and might have attributed this to divine concern for their well-being.[108]

Sociological studies of contemporary religious groups have documented the ways in which churches can attract members by offering companionship to lonely people.[109] Provided that a bundle of religious doctrines satisfies the needs of group members, promotes harmony within the group, and indirectly helps in generating new descendant groups or in taking over others, those ideas are likely to spread and may even become prevalent. Surely some religions have been very good at doing these things, at encouraging, for example, acts of great sacrifice to achieve religious rewards. We cannot yet aspire to tell the full story of why religions of so many different kinds have been prevalent across human societies, but the specific instances in which historical and sociological explanations can be given strongly suggest that the causes of success stem from the attractiveness of stories and alleged historical claims, on the emotions they provoke and the actions they inspire—and that they have nothing to do with the literal truth of those tales and histories.

* * *

Up to this point, the enlightenment case proceeds as if the religious believer had no direct access to sources of religious truth, but must rely on a tradition originating in the very distant past. Many religious people, however, think differently.[110] Christians talk of encounters with Jesus, and of an enduring presence in their lives. According to the statistics, religious experience is quite widespread—although perhaps the statistics are worrying, because the rates vary quite dramatically from year to year, decade to decade.[111]

I have no doubt that the overwhelming majority of the reports of religious experience are perfectly sincere. The important issue is why they occur. Religious people would prefer to think that the visionaries have, at least temporarily, a special ability to discern aspects of reality that ordinary experience can't disclose. The obvious scientific rivals invoke psychological and sociological causes—stimulation of normal sensory channels, against particular psychological backgrounds, induces people to assimilate their current experiences to the religious framework supplied by their culture, or by some culture with which they are familiar.

How can this issue be resolved? Not in the ways in which we corroborate other kinds of special powers—as when we test the musician who claims absolute pitch or the gourmet who is reputed to have an ability to detect the types and vintages of wines. Without an independent means of checking the believer's reports, it is hard to see how to reach any firm judgment. The point has been appreciated by religious groups, who have struggled to find ways of assessing self-described visionaries. The solution achieved in medieval procedures for certifying those with genuine religious experience was to compare their affirmations with the orthodoxies of church tradition. But, since religious experience is supposed to validate religious doctrine in a way that appeal to tradition cannot, that solution is inept as a resolution of the issue that confronts us.

Once again, the vast array of forms religious experience takes causes trouble. The visions of Jews, Muslims, and Christians differ in ways we might think of as fundamental until we attend to the reports offered by the Yoruba, the Inuit, and Australian aborigines about their own religious experiences. To propose that the religious experiences of those whose lives are full of encounters with goddesses, ancestors, and totemic spirits are to be understood in psychosocial terms, while those reported by Western monotheists are accurate representations of religious truths invites obvious and unpleasant questions about why the psychosocial explanation shouldn't be adopted more broadly. To maintain that all the religious experiences are completely correct is evidently impossible. The reports of visionaries are massively inconsistent. No religious believer thinks—or could think—that the central stories of all religions are literally true.

Once again, the obvious way to salvage some role for religious experience is to suggest that all sincere religious experiences disclose some aspect of the divine, but that this is overlaid and colored in each case by the social and psychological constructions of individuals and their rival cultural traditions. The Catholics who see the Virgin in a window in Brooklyn are enjoying a vision of something that transcends mundane reality, but they interpret it according to the specific ideas of their religious culture. The specific ideas cannot be recognized as literally correct—all we can say is that they, like their Yoruba counterparts, have been in touch with some "element of ultimacy."

A response like this cannot support supernaturalism—again, it points in the direction of spiritual religion. Supernaturalists cannot find it reassuring to be told that the idea of Jesus as a constant living presence is a psychosocial construction, even though the core of the experience is an accurate sense of the "transcendent." Moreover, what we know about the contexts in which religious experiences occur readily fosters a deeper skepticism. Troubled people, people whose emotional lives are disturbed, are significantly more likely to report religious experiences, and there are fragmentary suggestions that the administration of hallucinogens increases the rate at which such experiences occur.[112] It would be wrong to maintain that we know that sincere religious experiences are the products of delusion. We should recognize clearly that we don't know what to make of some parts of human experience. Given the extent of our ignorance in this area, supposing that religious experiences can somehow be assimilated to the categories and doctrines that have descended to us from ancient times is a blind leap.

Faith is frequently prized for its readiness to make that kind of leap. Many religious people would surely be impatient with the arguments I have rehearsed here, and would declare that the importance of religious commitment lies in the fact that it does not seek reasons. The proper religious attitude is one of trust. "So," a devout Christian, or Muslim, or Jew might declare, "I simply accept these claims about past events, these doctrines about what people should do and what they should aspire to be. To ask me to provide reasons—or to play clever games that try to show I have no reasons—is entirely beside the point."

Yet here the enlightenment case presses from a new direction. If you are going to use your religious attitudes to run your life, if you are going to let religious doctrine guide you to decisions that will affect the lives of others, then the willingness to leap without evidence, to commit yourself in the absence of reasons, deserves ethical scrutiny. As William Clifford, a late Victorian mathematician and apologist for science, saw very clearly, we do not usually endorse the behavior of people who act without reason, ardently convinced that things will turn out well. In Clifford's famous example, the ship owner whose wishful thinking leads him to send out an unsound ship is rightly held responsible when the passengers and crew drown. The earnest religious believer who supposes that God has commanded him to kill his son, or that religious doctrine requires him to eliminate the ungodly, or that it is wrong to undertake the operations doctors prescribe to save the lives of children, will subordinate ethical maxims he would otherwise use to guide his conduct to the dictates of faith, faith that is admittedly blind, supported by no defensible reason. We should protest that blind commitment, for, if it is allowed to issue in action, it is profoundly dangerous.

In practice, people don't protest, because they think of the religious doctrines that move them, and move their friends, as sources of a correct ethical attitude. If the enlightenment case, as developed so far, is cogent, they can have no basis for this judgment. It, too, is an article of blind faith. The true character of acting from unreasoned faith is revealed when you look at the actions of those who are moved by a different faith, at militant fanatics who aim to murder those who do not conform to their religion, for example. Christians will naturally think of themselves as different, but, as we have seen, there is no basis for holding that the religious doctrines they avow are any more likely to be correct than those of other faiths, even of radical and intolerant versions of other faiths. The blindness with which they commit themselves to acting in accordance with their preferred interpretation of a particular text is no different from that of people who would express a similar enthusiasm for the *Protocols of the Elders of Zion* or who would regard *Mein Kampf* as divinely inspired.

The same ethical mistake pervades all of these instances. Unreasoned acceptance is only tolerable if the religious attitudes adopted are so confined and restrained that they have no implications for consequential moral decisions. Blind faith requires a firm appreciation of the importance of not being earnest.[113]

The elements of the enlightenment case against supernaturalism are well established. They have been elaborated in considerable detail by many scholars during the past two centuries. I have been attempting to clarify the logical structure, and logical force, of these combined elements. My whirlwind tour through the enlightenment case aims to support two points. First, it shows the serious difficulties supernaturalists face when they try to invoke the unfathomable mystery of God's plan as a way of evading the apparently overwhelming evidence that the world in which we live was not designed by a providential Creator. Second, it enables us to understand why Darwin is the source of such vehement opposition, why he is seen as the chief villain in the promotion of atheism.

* * *

The line of argument I have developed throughout this essay shows Christianity in retreat. The evidence for an ancient earth compels us to say goodbye to Genesis, so that at least part of the Bible must be read as not literally true. (As I have noted above, this is an ancient and respectable approach to reading the scriptures.) Darwin's discovery of a single tree of life undercuts creationism, and requires that any action on the part of the deity must be remote. When we understand the messiness of the processes through which life unfolds, any design must be judged as largely unintelligent, any Creator as, at best, whimsical and capricious. Providential religion can only be sustained by supposing that God's design is an unfathomable mystery.

The attempted retreat of providentialism, the vague gesture toward unknowable purposes, can only be sustained if there's some ground for supposing that appearances are deceptive, that, behind the muddle of life, there is a Creator with deeper intentions. Any attempt to save providentialism must be committed to a specific piece of supernaturalist doctrine. Here, the enlightenment case exposes the troubles for supernaturalism that Darwinism brings into prominence, for it shows that there is no basis for holding that the received stories of this Creator are literally true. Parts of the enlightenment case are clearly separable from any Darwinian ideas, as with the critical reconstruction and analysis of scriptural texts. Yet there are good reasons why Darwin, not Wellhausen or Hume or Voltaire, is taken as the leader of the opposition to what is valuable and sacred.

For the enlightenment case is not widely appreciated, and most of the brilliant thinkers who have developed it are unread, if not unknown. More exactly, they tend to be unread and unknown in the United States. Adolescent students in European schools study some of the relevant figures, to a lesser extent in Britain, to a much greater extent in the countries of Western continental Europe.[114] American defenders of supernaturalist or providentialist religions, some of them literalists about Genesis, others literalists about significantly fewer of the scriptures, are protected from the shock of biblical criticism, of sociological history of religions, of anthropological studies that show the diversity of religious ideas, of psychological evidence about religious experience, and of ethical reflections on the dangers of unreasoned decisions. When these potentially dangerous ideas surface, they are dismissed with brusque denials—the strategy of the *King James Study Bible* might, rather harshly, be described as one of lying for God.[115] Not only are the individual pieces mostly unrecognized, but the enlightenment case is rarely presented as a whole, as I have developed it (albeit briefly) here.

Darwin, however, is visible. He is in the schools, potentially corrupting the youth and leading them to spurn the precious gift of faith. He serves as the obvious symbol of a larger attack on supernaturalist religion, about which thoughtful Christians know, even if they are not aware of all its details. Their concern is justified, although they may think, wrongly, that the onslaught on their faith is contained and condensed in Darwinism. For the enlightenment case will not surface in the education of their children, at least not until they attend universities, and probably not in any systematic way, even then. To defend the faith the important step is to keep Darwin out of the classroom, or, failing that, to "balance" his corrosive influence.

Intelligent design-ers, like the scientific creationists before them, promise a way to do just that. They raise sufficient dust about "unsolvable problems" for Darwinian evolution to give concerned people the hope that there is a genuine alternative, friendlier to faith and acceptable with good conscience. When these advertisements are probed, as I have probed them in previous chapters, they are found to be thoroughly false. Overwhelming evidence favors the apparently menacing claims of Darwinism. Worse still, the threat to providentialist and supernaturalist religions, forms of religion that are firmly entrenched in many contemporary societies, turns out to be genuine.

Where does this barrage of arguments leave us? Darwin's most militant defenders would insist that they take us all the way to secularism, even that they constitute a knockdown case for atheism. I dissent from that conclusion for two reasons. First, even though the enlightenment case demonstrates that, taken as literal truth, the stories and historical claims of all the religions about which we know are overwhelmingly likely to be mistaken, it does not follow that the world contains nothing beyond the entities envisaged by our current scientific picture of it. The history of inquiry shows that our horizons have often expanded to encompass things previously undreamed of in anyone's natural philosophy. Whether inquiry will ever disclose anything that can satisfy the religious impulse, that can merit the title of "transcendent," is itself doubtful, and we can be confident that, even if this remote possibility is realized, it will not approximate any of the stories our species has so far produced. It would be arrogant, however, to declare categorically that there is nothing that might answer to our vague conception of the transcendent—there is too much that we know that we do not yet know.

Second, and more importantly, the critique of providentialism and supernaturalism leaves open the possibility of what I have called "spiritual religion."[116] Each of the major Western monotheisms can generate a version of spiritual religion by giving up the literal truth of the stories contested by the enlightenment case.[117] How can this be done? I shall illustrate the possibility by using the example of Christianity.

Spiritual Christians abandon almost all the standard stories about the life of Jesus. They give up on the extraordinary birth, the miracles, the literal resurrection. What survive are the teachings, the precepts and parables, and the eventual journey to Jerusalem and the culminating moment of the Crucifixion. That moment of suffering and sacrifice is seen, not as the prelude to some triumphant return and the promise of eternal salvation—all that, to repeat, is literally false—but as a symbolic presentation of the importance of compassion and of love without limits. We are to recognize our own predicament, the human predicament, through the lens of the man on the cross.[118]

Spiritual Christians place the value of the stories of the scriptures not in their literal truth but in their deliverances for self-understanding, for improving ourselves and for shaping our attitudes and actions toward others. Yet spiritual Christianity—like spiritual Judaism or spiritual Islam—is vulnerable from two directions. To those who have grown up in a more substantial faith, who have not appreciated the force of the enlightenment case and who see no need to abandon supernatural religion, the spiritual version seems too attenuated to count as genuine religion at all. So, even though many contemporary Americans agree that large portions of scriptural texts should not be read literally, most of them do not completely abandon supernaturalism in favor of spiritual religion. They continue to affirm that a personal God made a covenant with the Jews or that Jesus literally rose from the dead.[119] Where spiritual religion is most clearly visible, in explicit denials that the Jews were chosen in any straightforward sense or in attempts to explain the natural events that lie behind the conflicting resurrection narratives of the Gospels, the content of the religion seems to consist of powerful ethical ideas and exemplars.[120]

From the other side, secular humanists will see spiritual religion as a last desperate attempt to claim a privilege for traditions whose credentials have been decisively refuted. Secularists can find value in the teachings of Jesus, inspiration in the image of the sacrifice on the cross—but also in ideas of the Torah or the Qur'an, in the sayings of the Buddha, in Socrates and Augustine, Kant and Dewey, Gandhi and Du Bois. Moreover, they can acknowledge the power of the stories, their ability to move and to inspire, while insisting that these are not unique to religious literature. Why not go all the way, to a cosmopolitan understanding of thought about what is valuable and worth achieving, a secular conception that celebrates the very best in the ideas and stories from many different traditions, some of them unquestionably secular?

Pressed from two flanks, spiritual religion can easily appear unstable. On one side it is liable to lapse from clearheaded acceptance of the enlightenment case and to topple back into supernaturalism. On the other, it may replace partiality to a particular tradition—Judaism or Christianity say—and metamorphose into a cosmopolitan secular humanism. For a secular humanist, like me, spiritual religion faces the challenge of providing more content than the exhortations to, and examples of, compassion and social justice that humanists enthusiastically endorse, without simultaneously reverting to supernaturalism.[121] Although I do not see how that challenge can be met, it is not clear how to circumscribe all possible responses to it—and thus to close the case against religion, period. The enlightenment case culminates in a (polite) request to the reflective people who go beyond supernaturalism to spiritual religion, to explain, as clearly as they can, what more they affirm that secular humanists cannot grant.

That, I suggest, is where reason leads us. But it cannot be—nor should it be—the end of my story.

* * *

For, though they speak with the tongues of men and of angels, the voices of reason, as they have sounded so far in this essay, should not expect to carry the day. The conclusion they draw deprives religious people of what they have taken to be their birthright. In its place, they offer a vision of a world without providence or purpose, and, however much they may celebrate the grand human adventure of understanding nature, that can only appear, by comparison, to be a mess of pottage. Often, the voices of reason I hear in contemporary discussions of religion are hectoring, almost exultant that comfort is being stripped away and faith undermined; frequently, they are without charity. And they are always without hope.

Religion is, and has been, central to the lives of most people who have ever lived. From what we know of the history of the growth and spread of particular creeds, its pervasiveness is understood in terms of the social purposes it serves, and nobody should expect it to disappear without a struggle, under the impact of what proclaims itself—accurately, I believe—as reason. For the benefits religion promises to the faithful are obvious, and obviously important, perhaps most plainly so when people experience deep distress. Darwin doesn't provide much consolation at a funeral.

Of course, secularism has its own revered figures, people who met personal tragedies without turning to illusory comforts. Hume faced his painful death stoically, persisting in his skepticism to the end. T. H. Huxley, Darwin's tireless champion, wracked with grief at the death of his four-year-old son, refused Charles Kingsley's proffered hope of a reunion in the hereafter. Perhaps these figures should serve as patterns for us all, admirable examples of intellectual integrity and courage that will not take refuge by turning away from the truth, by supposing, with the supernaturalists, that stories about life after death are literally true.

It is crushingly obvious, however, that those most excited by the secular vision—those who celebrate the honesty of spurning false comfort—are people who can feel themselves part of the process of discovery and disclosure that has shown the reality behind old illusions. Celebrations of the human accomplishment in fathoming nature's secrets are less likely to thrill those who have only a partial understanding of what has been accomplished, and who recognize that they will not contribute, even in the humblest way, to the continued progress of knowledge. Hume's and Huxley's heirs, like Richard Dawkins for example, preach eloquently to the choir, but thoughtful religious people will find their bracing message harsh and insensitive. How can these celebrants of secularism understand what many other people stand to lose if their arguments are correct? How can they expect those people to be grateful for the mess of pottage they offer?

Because such questions naturally arise, many people resist those arguments, hoping that they are incorrect or incomplete. They know that the case launched against their cherished beliefs is clever, but they are also tempted by the thought that the cleverness is flawed. If others, recognizably more sympathetic to their faith, can point however vaguely to potential faults, they will be grateful—and they will be disinclined to inspect too closely the gifts they are offered. So, again and again, they view Darwin as the enemy of what they hold most dear, and they resist Darwinism with whatever devices their apparently sympathetic allies can supply.

Christian resistance to Darwin rests on the genuine insight that life without God, in the sense of a Darwinian account of the natural world, really does mean life without God in a far more literal and unnerving sense. Even those who understand, and contribute to, the enlightenment case can find the resultant picture of the world, and our place in it, unbearable. William

James' arresting image of the high cliffs that surround a frozen lake, on which the ice is slowly melting, testifies to his own yearnings for some way of enlarging, or enriching, the scientific worldview he felt compelled to accept. In our own day, the religious scholar Elaine Pagels has provided a moving account of her similar needs.

Throughout her distinguished career, Pagels has explored the variant doctrines within early Christianity, showing with great lucidity and subtlety how the canonical texts of the New Testament represent a selection from a much more varied set of religious ideas. She recognizes, as Wellhausen did more than a century before her, that her work undercuts the thought that central claims of the orthodox documents are literally true. Her religious perspective aims to move "beyond belief," to a spiritual religion of seeking and individual discovery, one that can find inspiration in many Christian sayings and stories, as well as in the teachings of other religions. The Gospel of Thomas, left out of the canon in the interests of "Christian truth," strikes her as particularly suggestive in pursuing her own quest.[122]

She would not always have viewed her life this way, for, as she explains, there was a long period during which she did not attend church. Then, after a morning run the day after she learned that her infant son had a disease that would lead to a very early death, she paused in the vestibule of a New York church. "Standing in the back of that church, I recognized, uncomfortably, that I needed to be there. Here was a place to weep without imposing tears upon a child; and here was a heterogeneous community that had gathered to sing, to celebrate, to acknowledge common needs, and to deal with what we cannot control or imagine. Yet the celebration in progress spoke of hope; perhaps that is what made the presence of death bearable.

Before that time, I could only ward off what I had heard and felt the day before."[123] This poignant account contains, I believe, much that is deeply insightful.

Pagels found uplifting music. She met sympathetic people, willing to listen and to talk with her about important things. She discovered a place in which there was no need to hide her grief. She became part of a family, a "family that knows how to face death," as she puts it.[124] That family brought her comfort.

In the most obvious sense, she did not find hope—or so, at least, I believe. The celebration may have told the familiar, comforting, Christian stories. Yet, tempting though it might have been to brush aside the enlightenment case in the need for consolation, in such urgent need as Pagels surely had, the hope generated by taking those stories as literally true would have been illusory. To believe in the genuine possibility of a future that would bring her personal tragedy to a happy ending—to envisage a reunion in the hereafter, as Kingsley suggested to the grieving Huxley—would be self-deception.

The importance of Pagels' precise description of what occurred in the church, and of the perspective she develops in her book, lies, I suggest, in the genuine possibility of comfort without supernaturalist hope. When the soprano soloist sings the movement Brahms added at the last moment to his *German Requiem*, "I will comfort you as one whom his mother comforteth" (*"Ich will euch trösten, wie einen seine Mutter tröstet"*), the promise is literally false—there is no God who will wipe the tears from our eyes—but the music itself consoles. In deeper and more enduring ways, so do the love and sympathy of others, the support of a caring community.

There is a tendency for those who can accept life without God to pride themselves on their intellectual integrity. They, unlike the ostriches of the "booboisie," can face the facts without flinching. It is easy to think that the dominance of secular perspectives within universities, and in other places where highly educated people are found, is readily explained in terms of clear-headedness and tough-mindedness. These are people who can appreciate the force of the arguments, and who will not allow reason to be clouded by weak emotions. I doubt, however, that that is a complete account. Academics and scientists, as well as other professionals, can more easily sustain a sense of their lives as amounting to something, even in the absence of faithful service to God. Their lives are centered on work that is frequently significant and challenging, exciting and rewarding. Typically, they belong to communities in which serious issues can be openly discussed, in which there are readily available opportunities for the sharing of troubles and concerns. Even so, when unanticipated personal trouble strikes, the mechanisms for providing comfort may be quite inadequate.

Pagels' moving testimony is important to remind us of the need for comfort, and, in doing so, it opens a window into the lives of the people who most vehemently resist Darwin. They are typically not as lucky as the fortunate secularists who can affirm the enlightenment case, embrace life without God, and get on with their interesting work, their comfortable leisure pursuits, and their rewarding discussions with friends and colleagues. For many Americans, their churches, overwhelmingly supernaturalist, providentialist churches, not only provide a sense of hope, illusory to be sure, but also offer other mechanisms of comfort. They are places in which hearts can be opened, serious issues can be discussed, common ground with others can be explored, places in which there is real community, places in which people come to matter to one another—and thus come to matter to themselves. Without such places, what is left?

Moreover, for those who have been victims of injustice, who have found themselves stigmatized or marginalized by the secular state, cut off from its benefits and subjected to unfair burdens, religious gatherings can serve as occasions for focusing legitimate protest. From the Old Testament prophets to Martin Luther King, Jr., religious leaders have offered the poor and downtrodden opportunities to reclaim their rights. From the meeting houses that have broadcast the outcry of the urban poor to the liberal Catholic churches of Latin America, religion has provided a place in which individual sufferings can be united in a political movement. At their best, the religions of the world have championed the causes of the oppressed.

To resist Darwin, or the enlightenment case that looms behind him, is hardly unreasonable if what you would be left with is a drab, painful, and impoverished life. For people who are buffeted by the vicissitudes of the economy, or who are victimized by injustice, or who are scorned and vilified by the successful members of their societies, or whose work is tedious and unrewarding, people for whom material rewards are scanty or for whom the toys of consumer culture pall, for people who can unburden themselves most readily in religious settings and who find in their church a supportive community, above all for people who hope that their lives mean something, that their lives matter, the secular onslaught threatens to demolish almost everything. That is why the voices of reason are as sounding brass or as tinkling cymbals.

Writing in the 1920s, thoroughly aware that the enlightenment case had created a "crisis in religion," America's premier philosopher, John Dewey, argued for a new attitude to religion and the religious. We need, he suggested, outlets for the emotions that underlie religion, and this requires the emancipation of the religious life from the encumbrance of the dogmas of the churches, of their commitment to the literal truth of their favored stories. The task is to cultivate those attitudes that "lend deep and enduring support to the processes of living."[125] Dewey was, I believe, pointing to a position on which spiritual religion and secular humanism can converge, the former by erecting barriers against sliding back into supernaturalism, and embracing a cosmopolitan conception of the contribution of many different traditions to our understanding of the deepest questions about ourselves and our ideals, the latter by giving up its bracing recommendations to move beyond superstition, and by appreciating the genuine needs that stand behind religion. "It is the claim of religions that they effect this generic and enduring change in attitude. I should like to turn the statement around and say that whenever this change takes place there is a definitely religious attitude. It is not *a* religion that brings it about, but when it occurs, from whatever cause and by whatever means, there is a religious attitude and function."[126] At the beginning of the twenty-first century, we haven't achieved the broadening of the religious life Dewey envisaged.[127] For most Americans, the only occasions that cultivate the attitudes that support the processes of living are dominated by the doctrines of the traditional religions. If anything, the forms of Christianity that have been most successful in recruiting new members place heavy emphasis on the full acceptance of dogma, on literal interpretations of the canonical texts. Despite the demolition of the doctrines that Darwin and his enlightenment allies ought to have wrought, scriptural myths pervade many

American lives because we have found no replacements for the traditional ways of supporting the emotions and reflections essential to meaningful human existence.

None of this is to deny that religion, as it has been elaborated in the substantive stories of the major traditions, is also capable of doing enormous harm. The history of religions reveals not only the consolations of the afflicted and the legitimate protests of the downtrodden but also the fanatical intolerance that expresses itself in warfare and persecution, that divides families, cities, and nations, that forbids people to express their love as, and with whom, they choose. We should not forget the last part of the enlightenment case, and its proper repudiation of subordinating ethical reflection to blind faith. It is possible to appreciate the ways in which the religions human societies have developed have met genuine human needs, without forgetting that the myths they have elevated as inviolable dogma have often been destructive. As one of many examples, we might recall the verdict of the Jesus Seminar. Mark's imaginative fiction about Pilate and the Jewish mob has been the source of profound misery and harm. Dewey saw our situation clearly—the challenge is to find a way to respond to the human purposes religion serves without embracing the falsehoods, the potentially damaging falsehoods, of traditional religions. We need to make secular humanism responsive to our deepest impulses and needs, or to find, if you like, a cosmopolitan version of spiritual religion that will not collapse back into parochial supernaturalism.

If the issues were clearly understood, that would stand forth as our crisis in religion, expressed in the recurrent battles about secular knowledge, of which the disparaging of Darwin is the most evident example. Why is evolution still controversial in the United States, even though opposition to Darwin is viewed with surprise and disdain in virtually all the rest of the affluent world? I offer a speculative answer, one that only touches on some dimensions of the issue.[128] American life is often a highly competitive scramble for material goods, one in which many people do not fare well. The social evolution of cities, small towns, and suburbs has led to increasing atomization, with ever fewer opportunities for shared civic life. Unlike their counterparts in Western Europe, Americans are often unprotected against foreseeable misfortunes. When difficulties threaten, or when they strike, people have few opportunities to converse about their worries and fears. When the material rewards seem tawdry and unsatisfying, when consumer culture appears arid and empty, when people have no sense of why their lives matter, they lack places in which to air their thoughts to others, to engage in exploration of possibilities. Many Americans can turn only to the churches for the sense of community that addresses the insufficiencies in their lives. There are often no secular alternatives. For plenty of Americans, there is no counterpart to the neighborhood pub or the piazza.

The democracies that have most fully appreciated the enlightenment case, that have been most successful in the transition to secularism, are those in which there are social networks of support. Citizens are protected from the risk of severe poverty; they are provided opportunities for taking care of their health.[129] Above all, there is a sense of community life, secular spaces in which people gather, and in which they can talk about their hopes and aspirations, their anxieties and troubles. Gatherings of this sort can provide the occasion for discussions that bring people to see what matters to them, what makes their lives significant—or, perhaps, the experience of the gathering itself, independently of what is said, can give a sense of meaning, of mattering. All these forms of support might still prove inadequate to the most extreme shocks that can beset our lives—they might not answer to all the needs that Pagels felt, and found met by the church "family." Nevertheless, to look for such systems of support, to develop and extend them to meet our human needs, to offer people opportunities to confront more directly what their lives might mean and why they might matter, seems to me to be the best direction in which to find a solution for our religious crisis.

There are, I believe, two sides to our problem, one social and one intellectual. The intellectual aspect arises because the most obvious places in which people can seek answers to the question of what their lives mean and why they matter are places dominated by supernaturalist religion, or, secondarily, are literary and philosophical texts imbued with supernaturalist doctrine. Recent philosophy, especially, but not only, in the English-speaking world has found little time for larger questions about the meaning and value of human lives, and the theologies of spiritual religions are complex and not broadly accessible.[130] We are a long way from William James' forthright declaration, "The whole function of philosophy ought to be to find out what definite difference it will make to you and me, at definite instants of our life, if this world-formula or that world-formula be the true one."[131] Yet if philosophers since James have treated the questions of how "world-formulas" bear on human lives with distaste or with disdain, writers and artists have been less fastidious, exploring the possibilities for meaningful life in a world beyond supernaturalism. One way to do philosophy as James conceived it would be to explain and elaborate on literary and artistic insights.[132] The intellectual problem, then, is urgent, deep, often neglected by champions of secularism—but not, I think, hopelessly intractable. If, in these last pages, I have focused more on the social side of the problem, it is both because I believe that social preconditions need to be met before intellectual solutions to questions about life's significance can be properly appreciated and assessed, and also because a sense of community can itself bring reassurances about the value of human lives. Fortunate people, embedded in well-functioning communities, can feel, deeply and securely, that their lives matter, without interrogating why this is so.

I offer only the roughest sketch of a serious problem, one of which intelligent design is the latest symptom. The vehement opposition to Darwin results in large measure from the existence of a powerful case, one in which Darwin's ideas play a significant and highly visible role, against supernaturalism and providentialism, the most widespread forms of Christianity and other traditional Western religions, coupled with a recognition that endorsing that case would leave many lives impoverished and empty. With good reason, people refuse to sell their souls for Wales, to trade what they view as their birthright for a mess of pottage.

There is truth in Marx's dictum that religion, more precisely supernaturalist and providentialist religion, is the opium of the people, but the consumption should be seen as medical rather than recreational. The most ardent apostles of science and reason recommend immediate withdrawal of the drug—but they do not acknowledge the pain that would be left unpalliated, pain too intense for their stark atheism to be a viable solution. Genuine medicine is needed, and the proper treatment consists of showing how lives can matter. An essential component of it is to address the social shortcomings to which I have pointed. We should look more carefully at the causes of the pain, the harsh competitiveness of American life, the lack of buffers against serious ills, the atomization of society, the vapidity of much secular culture, and above all, the absence of real community. We should articulate, as clearly as can be done, the possible routes along which lives can find significance. In addressing these issues we may discover that the deliverances of reason can be honored without ignoring the most important human needs—and, going beyond supernaturalism, that we can live with Darwin, after all.

NOTES

PREFACE

1. Charles Kitcher, "Lawful Design: A New Standard for Evaluating Establishment Clause Challenges to School Science Curricula," *Columbia Journal of Law and Social Problems*, 39:4, 2006, 451–494.

TEXT

2. Henry Morris, *The Remarkable Birth of Planet Earth* (San Diego: Creation Life Publishers, 1972), 75.
3. Theodosius Dobzhansky, "Nothing in Biology Makes Sense Except in the Light of Evolution," *American Biology Teacher*, 35, 1973, 125–129.
4. Percival Davis and Dean H. Kenyon, *Of Pandas and People*, 2nd ed. (Dallas: Haughton Publishing Company, 2004), ix.
5. This is made clear by the decision in *Lemon v. Kurtzman*.
6. There are earlier sources for the style of argument, both in the Christian and Islamic traditions. See, for example, Thomas Aquinas' "Fifth Way" (*Sunma Theologiae*, part 1; readily available

in Paul Sigmund, ed., *Saint Thomas Aquinas on Politics and Ethics* [New York: Norton, 1988], 32) and Averroes' "On Proving God's Existence," in *Faith and Reason in Islam* (Oxford: Oneworld, 2001). Interestingly, one of Hume's protagonists, Cleanthes, provides an elegant and lucid version of the argument; *Dialogues Concerning Natural Religion*, part II (New York: Macmillan, 1986), 143.

7. This point was documented very clearly by Barbara Forrest in her testimony at the Dover trial. A transcript is available at http://www.talkorigins.org/faqs/dover/day6pm.html
8. Joshua 10:12–14.
9. Some intelligent design-ers also advocate broader cosmological claims. I shall not consider these here.
10. The point has been commonplace among philosophers of science ever since the lucid formulation by the French physicist-historian-philosopher Pierre Duhem. See *The Aim and Structure of Physical Theory*, English translation of a work originally published in French in 1906 (Princeton: Princeton University Press, 1956). Intelligent design-ers are happy to draw on this philosophical consensus; see, for example, Stephen C. Meyer "The Scientific Status of Intelligent Design," in *Science and Evidence for Design in the Universe* (San Francisco: Ignatius Press, 2000), 151–211.
11. For more on the problems of testability as a criterion for genuine science, see Philip Kitcher, *Abusing Science* (Cambridge, MA: MIT Press 1982), chap. 2. In formulating these points, I am grateful to Elliott Sober, who has emphasized the need for independently justified auxiliary principles in his own work on intelligent design.
12. I am grateful to Stephen Grover for reminding me of this.

13. I believe that many of the best arguments in the recent decision in the case of the Dover School Board can be viewed in these terms. In particular, as the judge recognized, Kenneth Miller's testimony made it apparent that the teaching of intelligent design would serve no secular purpose.
14. Strictly speaking, Darwin doesn't commit himself fully to just one tree of life. The last sentence of the *Origin* speaks of a "few forms" or "perhaps just one" into which life was "originally breathed." I read this as typical caution (and the use of the Pentateuchal phraseology is surely intended both to disarm opposition and to reassure Emma). Darwin's successors have interpreted him as hypothesizing a single tree of life, and, henceforth, so shall I.
15. As with his basic evolutionary thesis about the tree of life, Darwin is cautious about the power of natural selection. Although he takes natural selection to be the chief agent of evolutionary change, he does appeal to other causes (some of which contemporary Darwinians would reject).
16. My vague formulation here glosses over an area of continued debate. While those who explore particular groups of living things, including some who are skeptical about evolution, can often agree about how to divide those groups into species, there is much controversy over the issue of how to say, in general, what species are. For presentations of some of the rival positions, see Marc Ereshevsky, ed., *The Units of Evolution* (Cambridge, MA: MIT Press, 1992).
17. Michael Behe, probably the leading figure in the intelligent design movement, adopts anti-selectionism early in his influential book *Darwin's Black Box* (New York: Free Press, 1996), 5. Unfortunately, there is a certain amount of hedging in his

formulation, and, as we shall discover, some backsliding later in the book.

18. Critics of intelligent design may think this is too charitable, that there are no clear disavowals of novelty creationism within the movement, or even renunciations of Genesis creationism. Although it's important to recognize the possibility of backsliding, I shall begin by giving the intelligent design-ers the benefit of the doubt. (As Jerry Coyne has pointed out to me, intelligent design-ers often phrase their ideas in ways that suggest sympathy for ambitious—"unofficial"—forms of creationism.)
19. As I have discovered, some well-educated people find this statement incredible. They suppose that nobody takes all the (nonpoetic) parts of the Bible as literal truth. Their reaction is surely based on the fact that all the religious people they know adopt nonliteralist strategies of reading the scriptures. In fact, as any survey of evangelical Christian literature reveals, literalism is extremely important to many Christians. This is apparent not only in the books written in support of "scientific creationism" (Henry Morris, *Scientific Creationism* [San Diego: Creation Life Publishers, 1974], 244; D. C. C. Watson, *The Great Brain Robbery* [Chicago: Moody Press, 1976], 11, 13), but also in the *King James Study Bible* (Nashville, TN: Nelson, 1983). The *Study Bible* begins its section on interpretation by reminding the reader that "the Bible is God's infallible, inerrantly inspired Word" (p. xxiii), and concludes a note on the opening of Genesis with the declaration that "the biblical account of Creation clearly indicates that God created the world in six literal days" (p. 6). See also note 129 below.
20. Gilbert White, *A Natural History of Selborne*, letter XXXI (Oxford: Oxford University Press [World's Classics], 1951), 209.

21. *Proceedings of the Geological Society of London*, 1831, 313–314.
22. There is a brilliant cartoon that depicts the scene I have described—with Noah's ark serenely placed on the placid waters in the background. Unfortunately, I have lost the source.
23. For a penetrating study in the household economy of the ark, see Robert A. Moore, "The Impossible Voyage of Noah's Ark," *Creation/Evolution*, issue XI, 1983. Michael Weisberg has pointed out to me the difficulties in meeting the need to supply Noah's family and the assembled animals with fresh water—if the ark had carried enough fresh water for all its occupants to drink, it would probably have sunk.
24. The total is 1 (the original) + 2 (the two groups it produces) + 4 + . . . + 256 = 511.
25. Magnificent though the Grand Canyon is, there are numerous other places where the history of life on our planet becomes visible—as Steffi Lewis reminded me, many railway cuttings will do.
26. In fact, the practice of dating is a little more complicated than this, in that dates are often assigned directly to lavas and ashes that occur in conjunction with rock layers.
27. For further discussion of radiometric dating and scientific creationist attempts to question it, see Philip Kitcher, *Abusing Science*, 155–164. To appreciate the current state of the art, see Alan P. Dickin, *Radiogenic Isotope Geology*, 2nd ed. (Cambridge: Cambridge University Press, 2005).
28. Henry Morris, *Scientific Creationism* (San Diego: Creation Life Publishers, 1974), 247. I have amended Morris' own metaphor.
29. Darwin, *Origin*, 138.
30. The modification can easily be attributed to natural selection. Since eyes no longer have a function inside the caves, selection

will favor those variants that reassign the resources previously committed to the development of the visual system.

31. Darwin, *Origin*, 167. It's worth noting that Darwin treats the creationist alternative as a potential piece of science, and rejects it, not because of its invocation of a creative agent but because it is explanatorily bankrupt.
32. Darwin, *Origin*, 434. My preceding list of questions is drawn from *Origin*, 186, 339–341, 397–401.
33. Darwin, *Origin*, 435.
34. Darwin summarizes the paleontological problems in *Origin* 342–343, and devotes a whole section to the difficulty with complex organs, focusing in particular on the eye (*Origin*, 186–194).
35. Genetic similarity provides a more fundamental criterion for assessing relationships of ancestry and descent than do the similarities in anatomy and physiology on which Darwin and his immediate successors drew. The overwhelming majority of older attributions of relationship endure, even though there are occasional instances in which genetic analysis reveals that one organism is a closer relative of another than of a third that has traditionally been taken to be its closest kin. Moreover, even brute genetic similarity can prove misleading, in that pieces of genetic material can sometimes be acquired without deriving from ancestors—viruses and bacteria can spread DNA into organisms they infect. Molecular analysis of relationships can usually identify these isolated occurrences by focusing on the systematic similarities and differences among genomes.
36. See J. J. Yunis and O. Prakash, "The origin of man: A chromosomal pictorial legacy," *Science*, 215, 1982, 1525–30.
37. Far less probable is a scenario in which the common ancestor had 23 pairs, for that would require three independent events of chromosomes splitting at the same place.

38. The sequence data concern the common chimpanzee. It is overwhelmingly likely that results comparing human DNA with DNA from bonobos would reveal a similar kinship.
39. Our increased understanding of the mechanisms of aging in multicellular organisms provides a wonderful illustration of this. Because of potential errors in copying DNA, multicellular organisms need DNA repair enzymes. Copying can only proceed by clipping a small amount from the DNA to be copied. So, in the replication of chromosomes in multicellular organisms, the placing of genes at the ends would involve the loss of crucial pieces of DNA. Multicellular organisms solve the problem by adding a repetitive sequence at the ends of the chromosomes—the *telomere*—and shortening this doesn't matter, at least for a while. Moreover, the telomere protects the organism against overzealous activity on the part of DNA repair enzymes, that might join chromosomes together and make cell division impossible. Of course, once the telomere has been whittled away, the cell lineage is effectively finished. That is sad, but manageable, for the somatic cells of a multicellular organism—but more problematic for the germ-line cells. They can't afford to lose their full telomeres, for, if they did, the protection would diminish from generation to generation, until, relatively speedily, it vanished. Hence there has to be another repair device, an enzyme (*telomerase*) that restores telomeres in germ-line cells. You might think that it would be advantageous to have telomerase available in all stem-cell lineages, but the disastrous effects of this are apparent in the unrestrained growth of some cancers. So the solution is imperfect—multicellular organisms live with the programmed death of most cell lineages. Telomerase has to be confined to germ-cell lineages, but there are occasional mistakes in which it becomes accessible in somatic tissues and

allows for uncontrolled growth. This entire story shows the way in which the machinery present in unicellular organisms gives rise to the need for limiting and regulating that machinery in the different context of multicellular life. (I am indebted to Robert Pollack for conversation about these phenomena. An accessible brief account is given in his book, *The Missing Moment* [New York: Houghton Mifflin, 1999], 135–138.)

40. My argument here tacitly assumes that the designing force has particular aims, that it is directed toward creatures that don't contain unnecessary bits and pieces or problematic parts. An alternative would be to suppose that the creative agency is oriented differently. I shall consider this possibility shortly.
41. Davis and Kenyon, *Of Pandas and People*, 113.
42. Davis and Kenyon, *Of Pandas and People*, 137.
43. Davis and Kenyon, *Of Pandas and People*, 136.
44. In his testimony at the Dover trial, Michael Behe explicitly and repeatedly declined to advance any hypotheses about the power and direction of Intelligence. I shall be returning to the explanatory shortcomings of this silence in the next chapter.
45. Davis and Kenyon, *Of Pandas and People*, 37.
46. Darwin, *Origin*, 310–311.
47. Phillip Johnson, *Darwin on Trial* (Washington DC: Regnery, 1991), 79. Johnson's arguments are echoed in the school text *Of Pandas and People*, 100–107.
48. Johnson, *Darwin on Trial*, 75. As I was preparing the final version of this essay, a second jewel may have been added to the crown, with the discovery of the fossil remains of *Tiktaalik* an intermediate between fish and land-dwelling animals (*Nature*, 440, 2006, 757, 764).
49. Johnson, *Darwin on Trial*, 75.
50. Johnson, *Darwin on Trial*, 76.

51. The known exceptions, most famously the Burgess Shale, are extremely rare—so rare that we can take the probability to be effectively zero. (Here I am indebted to David Walker.)
52. Johnson, *Darwin on Trial*, 76. Again the school text echoes Johnson (*Of Pandas and People*, 100–101). The confusions I discuss in the text are also found in many novelty creationist discussions of the hominid fossil record (for example, in both *Darwin on Trial* and *Of Pandas and People*).
53. Francis Darwin, ed., *More Letters of Charles Darwin*, vol. 2 (London: John Murray, 1903), 379.
54. Henry Charles Fleeming Jenkin, "The Origin of Species," *North British Review*, June 1867; reprinted in David Hull, ed., *Darwin and his Critics* (Cambridge, MA: Harvard University Press, 1974), 303–344, quoted material from 315–316.
55. As David Helfand has pointed out to me, the estimates have become ever more precise: the currently accepted figure is 4.54 billion years and is taken to be correct to within 1 percent.
56. This is the title of R. A. Fisher's seminal monograph, one of the landmark achievements of twentieth-century evolutionary theory. Fisher and Sewall Wright showed how the integration of genetics and Darwinian natural selection could be articulated mathematically; Dobzhansky then applied their theoretical ideas to the study of natural populations.
57. But there are significant differences in the formation of the gametes across the entire range of living things. For an enlightening discussion of variation in this respect, see Leo Buss, *The Evolution of Individuality* (Princeton: Princeton University Press, 1987).
58. There are currently subtle issues about the extent to which mutation rates can be preferentially directed. For the sake of simplicity, I shall ignore these debates, since any greater

liberalism about evolutionary mechanisms will tend to help defenders of Darwin rather than champions of intelligent design.

59. See Peter R. Grant, *Ecology and Evolution of Darwin's Finches* (Princeton: Princeton University Press, 1986, 1999). Jonathan Weiner, *The Beak of the Finch* (New York: Vintage, 1994) provides a wonderfully accessible account of the Grants' accomplishments. Frank Sulloway, "Darwin's Finches: The Evolution of a Legend" in *Journal of the History of Biology*, 15, 1982, 1–53 is an extraordinary piece of historical detective work that demolishes the popular view that Darwin's observations of the finches on the Galapagos played a crucial role in awakening him to evolution. For a review of other studies of natural selection in the wild, and of the methods used to confirm selection in action, see John Endler, *Natural Selection in the Wild* (Princeton: Princeton University Press, 1986).
60. The Grants' work provides slightly more grounds for an optimistic assessment, in that, under swings of harsh drought and seasons of heavy rain, they were able to trace significant changes in the forms of finch beaks. For the sake of the argument here, however, I am supposing that the intelligent design-ers are demanding something far more dramatic even than this.
61. Francis Darwin, ed., *Life and Letters of Charles Darwin*, vol. 2 (London: John Murray, 1888), 352.
62. For a superbly accessible presentation of this research, see Richard Dawkins, *Climbing Mount Improbable* (New York: Norton, 1996). Another lucid, and concise, account by one of the scientists involved is Dan-Eric Nilsson, "Vision Optics and Evolution" in *Bioscience*, 39, 1989, 298–307.
63. Behe, *Darwin's Black Box*, 70. Behe uses the example of the bacterial flagellum as a parade case in many of his writings and

presentations. See, for example, "Design at the Foundation of Life" in *Science and Evidence for Design in the Universe* (San Francisco: Ignatius Press, 2000), 120 ff.

64. Behe, *Darwin's Black Box*, 73.
65. Perhaps not complete fantasy. The account I offer here is concordant with a recent review of the molecular details of the bacterial flagellum. See Howard C. Berg, "The Rotary Motor of Bacterial Flagella," *Annual Review of Biochemistry*, 72, 2003, 19–54. (For this reference, I'm indebted to Mel Simon.)
66. For the full-dress treatment of the design inference, see William Dembski, *The Design Inference* (Cambridge: Cambridge University Press, 1998). For an incisive rebuttal, see Brandon Fitelson, Christopher Stephens and Elliott Sober, "How Not to Detect Design" in Robert Pennock, ed., *Intelligent Design Creationism and its Critics* (Cambridge, MA: MIT Press, 2001).
67. Behe, *Darwin's Black Box*, 94.
68. For advice about these probabilistic considerations, I am grateful to Isaac Levi and Erick Weinberg.
69. Stuart Pullen, *Intelligent Design or Evolution? Why the Origin of Life and the Evolution of Molecular Knowledge Imply Design* (Raleigh, NC: Intelligent Design Books, 2005).
70. Pullen, *Intelligent Design or Evolution*, 96.
71. Pullen, *Intelligent Design or Evolution*, 102.
72. For two examples, see Stuart Kaufmann, *The Origins of Order* (New York: Oxford University Press, 1993)—the main ideas of which are presented less precisely but more accessibly in Kaufmann's *At Home in the Universe* (New York: Oxford University Press, 1995)—and Marc W. Kirschner and John C. Gerhart, *The Plausibility of Life* (New Haven: Yale University Press, 2005).
73. Adam Smith, *The Wealth of Nations* (New York: Modern Library Classics, 2000), 484–485. One possible interpretation of

Darwin's work is to view him as transferring ideas from British political economy to the natural world.

74. For some ingenious analyses, see Thomas Schelling, *Micromotives and Macro-behavior* (New York: Norton, 1978).
75. See Przemyslaw Prusinkiewicz and Aristid Lindenmayer, *The Algorithmic Beauty of Plants* (New York: Springer, 1990), Hans Meinhardt, *The Algorithmic Beauty of Sea Shells* (New York: Springer, 1998), J. D. Murray, *Mathematical Biology* (New York: Springer, 1989) chap. 15, and, more accessibly, Murray "How the Leopard got its Spots" in *Scientific American*, 258, 1988, 80–87. This work was inspired by a seminal essay by Alan Turing, published near the end of his life.
76. In *God, The Devil, and Darwin* (New York: Oxford University Press, 2004), Niall Shanks provides a thorough and lucid elaboration of this important point and of its implications for claims made on behalf of intelligent design.
77. Behe, *Darwin's Black Box*, 7.
78. Recent studies in Burkina Faso indicate that the allele is significantly more prevalent there. Earlier investigations of a variety of Bantu populations showed low frequencies of the *C* allele, and raised the puzzle I discuss in the text.
79. I abbreviate the explanation given by Alan Templeton, "Adaptation and the Integration of Evolutionary Forces" in Roger Milkman, *Perspectives on Evolution* (Sunderland, MA: Sinauer, 1982). Another example in which an apparently beneficial allele is unable to spread in human populations may occur in connection with resistance to AIDS. It is possible that the small number of people with HIV who do not develop full AIDS carry two copies of an allele that is typically disadvantageous when it occurs in single dose. But this example is less well understood than the case described by Templeton.

80. As mentioned in note 78, some recent studies of populations in Burkina Faso have found the *C* allele in higher frequencies than those discovered in the groups investigated earlier. Does this show that Intelligence is, at last, swinging into action? Not really. First, it's possible for the accidents of mating and survival to allow for the presence of enough people with the advantageous *CC* combination, so that the *C* allele can attain the threshold frequency. Second, given that the mutation that gives rise to the *C* allele is so readily available, we can only wonder why Intelligence has permitted the problematic *S* allele to spread its damaging effects among so many people. During the past few thousand years, many human beings, including many children, have died either because they lacked protection against malaria (they were *AA*) or through the crises of sickle-cell anemia. If Intelligence has started to solve this problem in Burkina Faso—but not in many other sub-Saharan populations—we can only wonder why it has permitted the botched *AS* solution to persist so long, why it has done so little, so late.
81. William A. Dembski, *The Design Revolution* (Downers Grove, IL: Intervarsity Press, 2004), 251.
82. Dembski, *The Design Revolution*, 251.
83. Here I am indebted to conversations with Kent Greenawalt.
84. In their public presentations and lectures, they sometimes provide more than a wink. Thus William Dembski has directly linked intelligent design to the Christian scriptures, claiming that "Intelligent Design is the Logos of John's Gospel in the language of information theory" (presentation in 1998 to the Millstatt Forum in Strasbourg).
85. For an excellent example, see Kenneth Miller, *Finding Darwin's God* (New York: Cliff Street Books, 1999).

86. Here I diverge from the approach offered in *Abusing Science*. The last chapter of that book, coauthored with Patricia Kitcher, tried to defend the conciliatory line that views Darwinism as no threat to religion. My reasons for divergence will be explained in the text below. It should be noted, however, that my coauthor was always skeptical about the prospects for reconciliation, and she should not be held responsible for the earlier errors that I recant here.
87. G. J. Romanes, *A Candid Examination of Theism* (Boston: Thomas and Andrews, 1878), 114.
88. William James, *The Varieties of Religious Experience* (Harmondsworth: Penguin, 1982), 141–142. I take James to be reacting to a form of scientific naturalism inspired by Darwinism, exemplified in the writings of T. H. Huxley, John Tindall, and William Kingdom Clifford (in response to whom James had already written his celebrated essay "The Will to Believe").
89. James, *The Varieties of Religious Experience*, 162.
90. There are also providentialist strands within Judaism, as many passages in the Hebrew Bible make clear, as well as in Islam (the Qur'an emphasizes that the earth has been planned to provide for its inhabitants). But providentialism is particularly pronounced in Christianity.
91. The point is lucidly and forcefully made by Richard Dawkins, "God's Utility Function" in *River Out of Eden* (New York: Basic Books, 1995), chap. 4.
92. Darwin, *Origin*, 472.
93. Fyodor Dostoevsky, *The Brothers Karamazov* (New York: Modern Library, 1995), 271–272.
94. Indeed, if the afterlife includes the possibility of eternal punishment—as in the traditional Christian idea of hell—there are other problems that I don't consider here. See David Lewis,

"Divine Evil," forthcoming in Louise Antony, ed., *Philosophers Without Gods*. (Oxford) (I should point out that I reconstructed this essay from an outline Lewis left at his untimely death—most of the words are mine, but the arguments are all his.)

95. Darwin, *Origin*, 167.
96. Hume, *Dialogues Concerning Natural Religion*, part XI, 204.
97. This is quite compatible, of course, with not believing that many of the statements in the Bible are literally true. You can hold that the Bible is literally correct in declaring that God, who created the world, is deeply concerned with the fate of each of His creatures, while also denying that Genesis (and much of the rest of the text) is literally true.
98. Luke 2:1–4. Here, and elsewhere, I cite the Authorized (King James) translation. This is somewhat less accurate, but infinitely more beautiful than subsequent versions. It is also the translation often used by evangelical opponents of Darwin.
99. Judaism did evolve in response to the Roman destruction of the Temple, but, because of the ambiguous role of Jesus' followers in the Roman-Jewish war, the development of Judaism was not along the lines envisaged in the early Jesus movements.
100. Robert W. Funk and the Jesus Seminar, *The Acts of Jesus* (San Francisco: Harper, 1998), 153.
101. Robert J. Miller, ed., *The Complete Gospels* (San Francisco: Harper, 1994). As I was completing the last revisions of this essay, there were reports of yet another addition to the Gospels that have survived, the discovery of the Gospel of Judas, with its strikingly revisionary account of the betrayal of Jesus.
102. Gospel of Thomas, 12; Miller, *The Complete Gospels*, 307.
103. See Elaine Pagels, *Beyond Belief* (New York: Random House, 2003). It is also worth noting the opening verses of Luke's Gospel, in which the existence of rival Gospels is presupposed

by the evangelist's announced intention of setting the record straight. The clash between John and Thomas is probably one instance of a general phenomenon.

104. The letter of resignation is quoted in Richard Elliott Friedman, *Who Wrote the Bible?* (New York: HarperCollins, 1987), 165. I should note that Friedman disagrees with Wellhausen's assessment of the impact of biblical criticism (while I endorse it); see Friedman, *Who Wrote the Bible*, 243.
105. *King James Study Bible* (Nashville, TN: Nelson, 1985), 1401.
106. Similarly, both in discussions of Isaiah 7:14 and Matthew 1:23, the *Study Bible* simply denies standard views about how to translate the Hebrew word *almah* (young girl), which the Septuagint mistranslates as *parthenos* (virgin). Evangelical Christians prefer the King James translation to the more scholarly (and more prosaic) recent versions in which the verse in Isaiah refers to a "young woman," and Matthew's imaginative interpretation is more evident. See, for example, the Revised Standard Version (New York: Meridian, 1974), 605 (OT) and 1 (NT).
107. Rodney Stark, *The Rise of Christianity* (San Francisco: Harper, 1997), chap. 5.
108. The hypothesis was originally proposed by William H. McNeill in *Plagues and Peoples* (New York: Doubleday, 1977), 108–109. It is developed further by Stark, *The Rise of Christianity*, chap. 4.
109. See John Lofland and Rodney Stark, "Becoming a World-Saver," *American Sociological Review*, 30, 1965, 862–875.
110. The idea of religious experience as widely available and serving as some basis for faith is relatively modern, in part a reaction to Enlightenment critiques of reliance on texts and traditions. In his authoritative study, *Religious Experience* (Berkeley: University of California Press, 1985), Wayne Proudfoot traces the modern conception to the writings of Friedrich Schleiermacher

at the turn of the nineteenth century. I am extremely grateful to Proudfoot for his advice on this and other topics.

111. See Benjamin Beit-Hallahmi and Michael Argyle, *The Psychology of Religious Behaviour, Belief, and Experience* (London: Routledge, 1997), chap. 5.
112. See the discussion of the "Marsh chapel miracle" in *The Psychology of Religious Behaviour, Belief, and Experience*, 85 ff.
113. I have articulated the argument summarized here in much greater detail in an examination of William James' attempts to respond to Clifford's challenge. See "A Pragmatist's Progress: The Varieties of William James' Strategies for Defending Religion" in Wayne Proudfoot, ed., *William James and a Science of Religions* (New York: Columbia University Press, 2004), 98–138.
114. I am grateful to Christopher Peacocke for reminding me of this.
115. I borrow this phrase from a friend, the late Bob Schadewald, a tireless critic of the deceptions practiced by "creation scientists" in the 1980s, who intended that his book on the subject should bear the title *Lying For God*. The phrase was also used by Judge Jones in the Dover trial.
116. There are other labels that might be used for the attitude I have in mind—proponents of "spiritual religion" might be called "seekers," or people committed to "liberal theology." For a discussion that recognizes a variety of religious attitudes, see Robert Wuthnow, *Christianity in the Twenty-first Century* (New York: Oxford University Press, 1993).
117. Even more straightforwardly, spiritual religion can be elaborated along lines found in Eastern religions, perhaps most obviously in some versions of Buddhism.
118. Here I am indebted to some valuable comments by Stephen Grover.

119. This is evident from the ethnographic accounts offered in sociological studies of American religion. Whether believers are committed to some established denomination, or whether they fashion their own eclectic mix from the religious traditions of the world, they almost invariably take some scriptural claims to be literally correct. See, for examples, Wade Roof, *Spiritual Marketplace* (Princeton: Princeton University Press, 1999), Robert Wuthnow, *After Heaven* (Berkeley: University of California Press, 1998), and Richard Cimino and Don Lattin, *Shopping for Faith* (San Francisco: Jossey-Bass, 1998).
120. A fascinating example of the struggle to understand the concept of Jewish "chosenness" in the wake of post-Enlightenment approaches to religion is provided by Arnold Eisen, *The Chosen People in America* (Bloomington: Indiana University Press, 2006). In Eisen's account, a more substantive reading of chosenness emerges from an enrichment of the theology of Judaism—in my terms, a retreat from spiritual religion to supernaturalism. A full-dress spiritual account of the resurrection is offered by John Shelby Spong in *Resurrection: Myth or Reality?* (San Francisco: Harper, 1994). Spong's brave attempt to treat the death of Jesus as a thoroughly natural event comes to its climax in chapter 19 ("But what really did happen?") where he gives a "speculative reconstruction" of the events behind the familiar Gospel stories.
121. One source of the idea that spiritual religion provides more than secular humanism may lie in the tendency to conceive secular humanism in its most militant—"scientistic"—versions. I trust that the position will emerge more clearly, and more sympathetically, in the following pages.
122. Elaine Pagels, *Beyond Belief* (New York: Random House, 2003).
123. Pagels, *Beyond Belief*, 4.
124. Pagels, *Beyond Belief*, 3.

125. John Dewey, *A Common Faith* (New Haven: Yale University Press, 1934), 15.
126. Dewey, *A Common Faith*, 17.
127. Dewey's influence was visible for a time in the reforms proposed by Mordecai Kaplan—not only in his books but also in his establishment of Jewish Community Centers. Arnold Eisen's *The Chosen People in America* provides a clear account of Kaplan's work, and of its connections to Dewey's ideas.
128. A more systematic account would plainly have to deal with American populism, and with other egalitarian (anti-authoritarian) trends in our social and cultural history. My aim here is merely to indicate some deeper sources of the opposition to Darwin, sources that are less well appreciated.
129. There is considerable discussion of the extent to which the world is becoming more secular. Some scholars see a trend toward secularization in the affluent world, with the United States as a glaring exception. Others, including former advocates of the idea of a secular trend, suppose that religion is resurgent worldwide, and that it is the states of Western Europe that are exceptional. I take no stand in this debate, but focus solely on the acknowledged difference between the United States and Western Europe. Like some political scientists—for example, Pippa Norris and Ronald Inglehart, *Sacred and Secular* (Cambridge: Cambridge University Press, 2004)—I see vulnerability to risk as playing an important role in attracting people to religion. As Norris and Inglehart show with great lucidity and thoroughness, data from the World Values Survey and the European Values Survey support the hypothesis that people subjected to "existential insecurity" are much more likely to embrace religion. My own—admittedly speculative—account also adds an idea expressed by Peter Berger (in his

introduction to Berger, ed., *The Desecularization of the World* [Washington, DC: Ethics and Public Policy Center, 1999], 13, 4), when he declares, "The religious impulse, the quest for meaning that transcends the restricted space of empirical existence in this world, has been a perennial feature of humanity." Berger also points out that the contemporary manifestations of this impulse are not in the forms of religion carefully discussed by academics and intellectuals, but in "religious movements with beliefs and practices dripping with reactionary supernaturalism (the kind utterly beyond the pale at self-respecting faculty parties)."

130. For humanists, the evocative, but vague terminology of post-supernaturalist theology—"element of ultimacy," "object of ultimate concern," and the like—are signs of instability. Spiritual religion wants more than the ethical attitudes of secular humanism, but hopes to avoid supernaturalism, and the consequence often seems to be a lack of precision that has moved skeptical philosophers to make charges of meaninglessness. As I have suggested above, I prefer to pose the challenge of saying how a religion that rejects supernaturalism can be more than secular humanism with some special phraseology—or, even better, to recast the issue of the religious life as Dewey did.
131. William James, *Pragmatism* (reprinted with *The Meaning of Truth* [Cambridge, MA: Harvard University Press, 1978]), 30.
132. I have tried to make this clear in the cases of Wagner and Joyce. See Philip Kitcher and Richard Schacht, *Finding an Ending: Reflections on Wagner's "Ring"* (New York: Oxford University Press, 2004), and Philip Kitcher, *Joyce's Kaleidoscope: An Invitation to "Finnegans Wake"* (New York: Oxford University Press, forthcoming). There are many other literary and artistic sources on which similar work could be done.

INDEX